Breaking into IT

Your Roadmap to Success in the Tech Industry

Tyler L Strong, Sr.

ISBN's

Paperback - 978-1-964031-00-2
eBook - 978-1-964031-01-9
Audio - 978-1-964031-02-6

Table of Contents

Breaking into IT: Your Roadmap to Success in the Tech Industry

Preface

My journey into the tech industry was initially defined by roles at two telecommunications firms, spanning over five years. This period was foundational, setting the stage for a deep-seated passion for information technology. The moment I encountered the concept of IT certifications, a new path unfolded before me, marked by both challenges and discoveries.

The trail towards achieving the CompTIA Network+ certification was my first real test of perseverance in the IT field. After a disappointing failure on my initial attempt, I regrouped and changed my study approach, emphasizing understanding over memorization. This strategic shift paid off when I passed the exam, affirming the critical lesson that resilience is key to over-coming obstacles.

Inspiration from a mentor led me to tackle the Cisco Certified Networking Associate (CCNA) certification next. Achieving this not only broadened my technical expertise but also highlighted the profound impact of guidance and ambition in reaching my career objectives. However, despite these achievements, the networking positions I eagerly sought remained out of reach. This unforeseen challenge forced me to reassess my career trajectory. With an open mind and a willingness to adapt, I transitioned into the cyber-security domain. This shift was more than a mere change of focus; it was a leap into an area that required quick mastery of new skills and adaptation to its distinctive challenges.

This was the beginning of an intense phase of personal and professional growth. Embracing this new challenge, I immersed myself in the world of cyber, determined to adapt and excel. This determination was fueled by an unwavering work ethic and a deep commitment to mastering the skills required in this dynamic field.

Starting as a Security Operations Center (SOC) Analyst, it felt like I was drinking through a fire hose of information. I quickly recognized the importance of continuous learning and certifications for staying ahead. Driven by a relentless pursuit of excellence, I embarked on a rigorous journey to obtain relevant cybersecurity certifications. This period was marked by long hours of study, hands-on practice, and a disciplined approach to learning.

My efforts culminated in achieving several key certifications each representing a milestone in my growing expertise. These achievements were not mere accolades but proof of my dedication to becoming a proficient cybersecurity professional. They stressed my belief in the power of hard work and the importance of staying updated with the latest knowledge and trends in the field.

Motivated by my own experiences of overcoming hurdles and the pivotal role mentorship played in my journey, I founded The GRID. This initiative is a testament to the lessons learned along the way and my commitment to giving back to the IT community. The GRID aims to offer support, resources, and a network for IT professionals at any stage of their career, providing a foundation for success and growth in the ever-evolving tech industry.

As you engage with this book, I hope my story serves as a beacon of inspiration. May it encourage you to navigate the complexities of IT with determination, seek out mentorship, and seize the myriad of opportunities the field offers. At The GRID, success stories prosper, and yours could be next. Learn more and become part of our story at www.unitethegrid.com. If you find this book helpful along your journey, I encourage you to return to the platform where you obtained the book to leave a good review. This will help others find the book, allowing it to help a wider audience.

Chapter 1: The Foundation: Understanding the IT Landscape

In the changing world of Information Technology (IT), the possibilities are as expansive as the digital realm itself. From the intricate dance of code in software development to the vigilant fight against cyber threats in cybersecurity, IT unfolds as a vast tapestry woven with diverse threads of expertise. As we embark on this journey into the heart of the industry, we find ourselves at the crossroads of innovation, where technology intertwines seamlessly with the needs of our modern society. Welcome to a world where each keystroke, each line of code, and each network connection contributes to the ever-expanding frontier of IT.

This chapter will serve as a compass, guiding us through the diverse landscapes of software development, network administration, cybersecurity, data science, and the myriad of roles that collectively form the IT ecosystem. It's not just about gaining knowledge, but about nurturing the ability to grow and evolve with technological currents. This book will not only lay down the foundations for you but will also be your guide to building a career that withstands the test of time. In the coming chapters, we'll explore how to cultivate a mindset of continuous learning and adaptation, ensuring that your skills remain on the cutting edge, and your career trajectory soars in the dynamic world of Information Technology.

Essential IT skills and knowledge areas

In the constantly shifting world of IT, the role of cybersecurity stands as a beacon, ensuring the integrity and security of digital spaces. Cybersecurity is the art of protecting digital places, a calling for those drawn to the strategic dance of ethical hacking, risk assessment, and digital defense. Key field areas include:

- **Ethical Hacking:** This involves simulating cyber-attacks to identify vulnerabilities in a system. Ethical hackers use the same tools and techniques as malicious hackers, but legally and with the aim of improving security before actual breaches occur.
- **Risk Assessment:** It's about identifying, analyzing, and prioritizing risks to a system's security. This process

includes evaluating potential threats, determining their likelihood and impact, and then developing strategies to mitigate these risks effectively.

- **Incident Response:** This field involves handling security breaches or attacks efficiently. It includes identifying and containing the breach, eradicating the threat, recovering systems, and learning from the incident to prevent future vulnerabilities.
- **Threat Intelligence:** This involves collecting and analyzing information about emerging threats and attack methods. This intelligence guides proactive security measures, helping organizations stay ahead of potential security breaches.
- **Security Architecture:** The focus on designing and maintaining a secure infrastructure. It involves creating a comprehensive security framework that encompasses hardware, software, policies, and procedures to protect against threats.
- **Digital Forensics:** This field is about investigating and analyzing digital data following cybercrimes. It involves uncovering how a breach occurred, identifying the perpetrators, and collecting evidence that can be used in legal proceedings.

Cybersecurity experts are not just reactive to threats; they are the architects of robust security systems. They continually adapt their strategies, integrating innovative solutions like artificial intelligence and machine learning to predict and counteract sophisticated cyber threats. Their role involves anticipating potential vulnerabilities and employing a strategic approach to prevent breaches. Staying updated on the latest cybersecurity trends and technologies is not just a job requirement; it's a passion that drives these specialists to excel in their field.

In this digital age, cybersecurity professionals play an important role in shaping policies and frameworks that govern internet safety, while ensuring a balance between technological advancement and ethical governance. The ethical landscape in cybersecurity is complex and ever-evolving. Professionals in this field must constantly balance the need to protect against the imperative to respect user privacy and data integrity. They operate at the intersection of technology and ethics, ensuring their work not only defends against threats but also aligns with broader societal values.

Collaboration is a cornerstone in the world of cybersecurity. Professionals actively engage in knowledge-sharing forums, participate in joint efforts to combat threats, and often work in teams. Their collaborative efforts often extend into international territories, as cyber threats know no borders, making global cooperation and unified cybersecurity protocols mandatory to the field's success. This collective spirit extends beyond mere information exchange; it fosters a sense of community and shared responsibility in safeguarding our digital future. Information sharing, threat intelligence analysis, and collaborative responses to incidents characterize the interconnected web of cybersecurity experts striving to keep the digital frontier secure. In this symphony of vigilance, each cybersecurity professional plays a unique role, contributing to the harmony that protects the foundations of our digital world.

Data Science: Unveiling Insights in a Sea of Data

Data Science stands at the forefront of our digital age, transforming raw data into meaningful insights. It is the science and art of extracting knowledge from data, applying a blend of statistical analysis, machine learning, and data visualization techniques. This field is not just about handling numbers; it's about making data tell a story. Let's look at key skills:

- **Statistical Analysis:** This skill is fundamental in data science, enabling specialists to make sense of complex data sets. It involves using statistical methods to collect, review, analyze, and draw conclusions from data. This skill is key for identifying trends, testing hypotheses, and making data-driven decisions.
- **Machine Learning:** Data scientists use a method called machine learning to create programs that can predict outcomes or make decisions. Imagine teaching a computer to recognize patterns and learn from them, much like how we learn from experience. This involves using different types of machine learning methods. Some are simple, like plotting a straight line through data points (linear regression), while others are complex, like mimicking how the human brain works (neural networks). The key is choosing the right

method for the right task, so the computer can effectively learn from the data it's given.

- **Data Wrangling:** Often, data is not in a clean, ready-to-analyze state. Data wrangling, or cleaning and structuring raw data, is the step before any analysis. It involves dealing with inconsistencies, missing values, and transforming data into a more usable format.
- **Programming Proficiency:** Programming expertise in languages like Python or R is valued. These languages offer libraries and frameworks specifically designed for data analysis, making tasks like data manipulation, statistical analysis, and visualization more efficient.
- **Data Visualization:** The ability to translate complex data findings to understandable visual representations. This skill involves using tools and techniques to create charts, graphs, and dashboards that effectively communicate the results of data analysis.
- **Big Data Technologies:** Knowledge of big data tech is important for handling large datasets that cannot be processed using traditional data processing techniques. Tools like Apache Hadoop or Spark are used for distributed storage and processing of big data sets.

Data science begins with transforming raw, often disorganized data into a structured, understandable format. This process, known as data wrangling, is made efficient through programming skills in languages like Python. These tools help data scientists analyze and model data, revealing insights that inform decisions. The foundation of data science also lies in a strong understanding of statistics, necessary for interpreting data and making predictions. Machine learning uses historical data to forecast future trends, making these specialists highly desired in various industries.

The essence of data science is a blend of technical prowess and a mindset driven by curiosity and exploration. Data scientists, often seen as modern navigators in a sea of information, use their skills in data visualization, programming, and statistics to bring data stories to life. Their work isn't just about analyzing data; it's about extracting meaningful insights for informed decision-making. This field is constantly changing, presenting endless opportunities for discovery and innovation, making it an exciting and dynamic domain for those intrigued by data's hidden narratives.

Software Development: Crafting the Digital Symphony

Software development is the heartbeat of technological innovation, turning ideas into reality through coding. It's a field that combines creativity with logic, where developers craft everything from simple applications to complex systems. In this world, code is more than a language; it's the medium through which problems are solved and innovations are born. Core skills:

- **Programming Language Proficiency:** The knowledge of languages like Python, Java, or C++ is invaluable. It involves not just writing code, but understanding how to structure it, troubleshoot, and optimize for performance.
- **Understanding Algorithms & Data Structures:** This is about the logic behind the code. Efficient algorithms and data structures are crucial for developing software that is not only functional but also efficient and scalable.
- **Version Control:** Mastery of version control, especially tools like Git, is necessary. It's about keeping track of changes in the codebase, enabling collaboration, and maintaining a history of the project's evolution.
- **Knowledge of Databases:** Understanding database management, whether it's SQL or NoSQL databases, is important. It involves knowing how to store, retrieve, and manipulate data effectively.
- **Front-End and Back-End Development:** Skills in both front-end (creating the user interface and experience) and back-end (working on the server, database, and application logic) are treasured. Full-stack development knowledge, covering both areas, is particularly sought after.
- **Debugging and Testing:** Ability to identify and fix bugs is as important as writing code. Testing ensures the soft-ware works as intended and is free of defects.

Software development merges the artistry of coding with the practicality of problem-solving. It's more than just writing code; it's about crafting efficient, scalable solutions, understanding various development methodologies, and effective teamwork. Developers must not only be proficient in programming languages but also adept in software architecture, design patterns, and version control systems. This field spans various specializations, each contributing uniquely to our digital world, from front-end aesthetics to back-end functionality.

Software developers find joy in creating robust, maintainable applications. This sector is a blend of creativity and technical skills, transforming coding into an opportunity for innovation. As the digital world evolves, software development offers a dynamic, fulfilling career path for those passionate about making an impact through technology, maintaining its appeal as a diverse and exciting field.

Network Administration: Architecting the Digital Infrastructure

Network administration is pivotal in the IT ecosystem, focusing on the reliability and efficiency of network infrastructure. It's the art of ensuring seamless communication and connectivity within an organization. Network administrators are the unsung heroes who manage and maintain the complex networks that businesses rely on. Main skills include:

- **Network Design and Implementation:** Crafting and setting up networks to meet organizational needs. This includes understanding different networking topologies, protocols, and deploying hardware like routers and switches.
- **Network Security:** Protecting network infrastructure from cyber threats. This involves implementing security measures like firewalls, intrusion detection systems, and ensuring network compliance with security policies.
- **Troubleshooting and Problem Solving:** Quickly identifying and resolving network issues. Network administrators must be proficient at diagnosing problems and implementing solutions to restore functionality.
- **System Administration:** Managing the daily operations of computer networks. This includes tasks like setting up user accounts, managing permissions, and ensuring the systems are up-to-date.
- **Network Monitoring and Analysis:** Constantly monitoring network performance. This involves using tools to track network health, analyzing data for potential issues, and optimizing performance.

- **Communication Skills:** When explaining technical details to non-technical staff and collaborating with other IT experts, effective communication is imperative.

Network administration is important to the flow of data across networks, requiring a blend of meticulous problem-solving and managing complex network infrastructures. Beyond hardware setup, it involves crafting efficient systems, understanding protocols like TCP/IP, DNS, DHCP, and implementing security measures against network threats. Network administrators engage in proactive troubleshooting, requiring practical hardware expertise to apply theoretical knowledge effectively. They handle diverse network issues, emphasizing both immediate resolutions and performance optimizations.

In this field, technical skills combine with strategic foresight. Administrators must keep pace with evolving technologies and security trends to ensure network integrity. Their role extends beyond maintaining current functionalities to strategizing for long-term network resilience. As custodians of the digital trails connecting our world, network administrators play a huge role in maintaining the reliability and security of our increasingly interconnected society.

IT Project Management: Navigating Digital Initiatives

IT Project Management focuses on the planning, execution, and delivery of projects. It's a field where strategic planning meets technological implementation. Project Managers are the orchestrators who ensure that projects align with business goals, are delivered on time, and stay within budget. Core skills include:

- **Project Planning and Scheduling:** Developing detailed project plans, including timelines, milestones, and resource allocation. This involves understanding project requirements and setting realistic schedules.
- **Budget Management:** Keeping projects within financial constraints. It's about allocating resources efficiently and managing costs throughout the project lifecycle.
- **Risk Management:** Identifying potential project risks and developing strategies to mitigate them. This is necessary for ensuring projects proceed smoothly despite uncertainties.

- **Team Leadership and Communication:** Leading and motivating project teams, facilitating communication among team members, and acting as a liaison between technical teams and stakeholders.
- **Quality Assurance and Testing:** Ensuring that project deliverables meet quality standards and function as intended. This involves overseeing testing processes and addressing any quality issues.
- **Stakeholder Management:** Effectively communicating with and managing expectations of all parties involved in a project, including clients, team members, and management.

In the realm of Information Technology, Project Management plays a key role, blending strategic planning with technical implementation. It involves not just orchestrating the various phases of a project but also aligning them with business objectives, ensuring timely delivery within budget. Project Managers employ a range of skills, from meticulous project planning and budget management to effective team leadership and stakeholder communication. They also focus on risk management and quality assurance to ensure project success.

Central to this discipline is the capacity to balance technical expertise with business acumen. Project Managers excel in adapting various methodologies like Agile and Waterfall to suit project needs, while also managing team dynamics and stakeholder expectations. They are responsible for translating complex project aspects into understandable terms for all parties, fostering a collaborative environment.

Cloud Computing: Transforming IT Infrastructure

Cloud Computing has revolutionized how businesses approach IT infrastructure. It allows organizations to use computing resources over the internet, offering flexibility, scalability, and cost-efficiency. This paradigm shift has made cloud computing an essential aspect of modern IT strategies. Let's look at key skills:

- **Understanding of Cloud Service Models:** Familiarity with models like Infrastructure as a Service (IaaS), Platform as a Service (PaaS), Software as a Service (SaaS), and the differences between them. This involves knowing how to leverage each model based on project needs.

- **Cloud Migration and Deployment:** Skills in migrating existing infrastructure and applications to the cloud. This includes planning, executing, and managing the migration process efficiently.
- **Cloud Security:** Ensuring data security and compliance within the cloud. This involves understanding cloud specific security concerns and implementing robust security measures.
- **Cloud Infrastructure Management:** Managing and optimizing cloud resources. This includes monitoring performance, managing costs, and ensuring scalability and reliability.
- **Knowledge of Cloud Providers and Platforms:** Proficiency with platforms like AWS, Azure, or Google Cloud. Understanding the unique offerings and tools of each platform is recommended.
- **Automation and Orchestration:** Automating repetitive tasks and orchestrating complex workflows in the cloud. This skill is for improving efficiency and reliability.

Cloud computing, a core driver in IT innovation, transforms traditional technology infrastructure with its scalability and flexibility. This paradigm shift allows organizations to efficiently access resources like servers and storage over the internet, adapting easily to their varying needs. It democratizes advanced computing, offering small and large businesses alike the tools to streamline their operations without heavy upfront hardware investments.

This field's diverse models, including public and private clouds, provide customized solutions to meet unique business requirements. For future technological trends like AI and big data, cloud computing is a foundational platform for innovation. As we probe deeper into IT, it's about aligning your skills and interests with the myriad of opportunities in this field, guiding you towards a fulfilling career path in the diverse and dynamic world of IT.

Identifying Personal Interests and Strengths

When identifying personal interests and strengths, it's important to understand where they lie. This self-awareness helps to carve out a fulfilling career in IT. The following self-assessment is designed to help you reflect on your unique traits and how they align with various roles in the industry. There are no right or wrong answers here, only insights into your personal preferences and abilities.

Self-Assessment Questions

1. **Problem-Solving**: Think about a time when you successfully solved a difficult problem. What was your approach? Did you enjoy the process of finding a solution?
2. **Creativity and Innovation**: Reflect on activities where you feel most creative. Do you find excitement in coming up with innovative ideas or solutions?
3. **Attention to Detail**: Recall instances where your attention to detail was exceptional. How do you ensure accuracy and precision in your tasks?
4. **Communication Skills**: Consider your communication style. Are you comfortable explaining complex ideas in a simple way? Do you enjoy collaborative discussions?
5. **Adaptability**: Think about a situation where you had to adapt to a significant change. How did you handle it? What did you learn from that experience?
6. **Teamwork**: Reflect on a team project or activity you were part of. What role did you play? Did you enjoy working as part of a team?
7. **Tech Curiosity**: What recent technological advancement have caught your interest? How do you approach learning about new technologies?

Reflection and Exploration

After answering these questions, take some time to reflect on your responses. Do you see patterns or preferences emerging? The next section of this book will discuss how these personal attributes can guide you to find your niche in the field. By understanding your strengths and interests, you can better navigate the diverse and dynamic field of IT, finding a path that not only suits your skills but also ignites your passion.

Analysis of Self-Assessment Responses

Following the self-assessment, let's dive into understanding what your responses may indicate about your potential fit in the IT world. This analysis can help you align your strengths and interests with suitable career paths:

- **Problem-Solving**: If you enjoyed and excelled in problem-solving scenarios, careers in areas like software development, network administration, or cybersecurity might be appealing. These fields require a knack for identifying issues and devising effective solutions.
- **Creativity and Innovation**: A fondness for creativity and innovation suggests you might thrive in roles like UI/UX design, game development, or emerging tech fields. These areas value fresh ideas and creative problem-solving.
- **Attention to Detail**: If your strength lies in meticulous attention to detail, consider roles in quality assurance, database management, or data analysis, where precision is invaluable.
- **Communication Skills**: Strong communicators often do well in project management, technical consulting, or customer support roles, where explaining complex concepts is a frequent occurrence.
- **Adaptability**: If adaptability stood out in your responses, look towards fields that are constantly evolving, like cloud computing or IT consultancy, which require a willingness to continuously learn and adapt.
- **Teamwork**: A preference for teamwork might lead you to collaborative environments like agile software development teams or IT project groups.
- **Tech Curiosity**: If a specific technology intrigued you, explore specialized fields related to that technology. For instance, an interest in AI could lead to machine learning roles.

This guide is designed to help you explore IT careers that match your traits, encouraging you to probe deeper into these roles for informed career decisions. Researching through videos or blogs, especially 'day in the life' content on platforms like YouTube, offers insights into various roles. Reflect on your self-assessment outcomes and the potential career paths aligned with your strengths and interests. The next steps involve translating these insights into actionable plans, exploring the IT world in more detail through each chapter. This journey is about connecting your traits to practical career opportunities, setting the stage for personal and professional growth in the field.

The path in IT is not linear; it's a landscape of opportunities where your unique blend of skills and interests can find a place. Embark on this exciting journey with an open mind and a willingness to explore. The field of Information Technology offers endless opportunities for growth, innovation, and fulfillment. Your unique combination of skills and passions can lead to a rewarding career in this dynamic industry. Stay curious, stay motivated, and let your journey into the tech industry begin!

Reflect and Apply Questions

1. **Discovering IT's Impact:** Reflect on how different aspects of Information Technology influence your daily activities. Can you identify moments where IT plays a hidden but crucial role in your routine?
2. **Personal Technology Analysis:** Think about your favorite piece of technology (like a smartphone or laptop). What features or functionalities fascinate you the most, and why do you think they are important in the broader context of IT?
3. **Career Inspiration:** After learning about the diverse areas within this industry, which field interests you the most, and what initial steps might you take to explore that interest further?
4. **Technology in Society:** How do you see technology shaping society in the next decade? What role do you think IT professionals will play in this transformation?
5. **Learning and Growth:** As someone new to the field, what are the skills or knowledge areas you believe are essential to start learning, and how do you plan to acquire them?

Chapter 2: Current and Emerging Technologies

Emerging technologies, the driving force behind modern IT innovations, offer a plethora of opportunities for those entering the field. These technologies, evolving rapidly, are reshaping industries and societal norms, extending from AI and blockchain to IoT and cloud computing. For aspiring IT professionals, understanding these technologies is vital for identifying future career paths and areas of growth. Each technology presents unique opportunities beyond their traditional applications, whether it's AI's role in data analysis and ethical considerations, blockchain's applications in secure transactions, or IoT's impact on smart technology. This chapter aims to clarify these technologies, highlighting their potential and the diverse skills required to harness their power, thus paving the way for a dynamic and fulfilling career.

As we explore these technologies, it's important to recognize the rapid pace at which this field is evolving. To truly grasp and stay ahead of these changes, consistent individual research and diligent note-taking are invaluable. This approach not only helps in understanding the current state of the technology but also prepares you to adapt to future advancements. Keeping a dedicated notebook or digital document for your observations, questions, and insights can be immensely beneficial. This practice will enhance your learning experience, making it more engaging and personalized. This is also an excellent practice to help you excel in your career. At times, IT jobs can feel like drinking from a firehose of information. Being able to take detailed, organized notes will benefit your ability to learn by having an easily accessible reference.

Foundational Technologies

In this section, we'll discuss the fundamental pillars of IT infrastructure, exploring the core principles that underpin modern technological advancements. From understanding the basics of this infrastructure to exploring the immersive realms of virtual and augmented reality, you will gain insights into the building blocks of today's digital landscape.

Basics of IT Infrastructure

Before we talk about current and emerging technologies, lets discuss the basics, like infrastructure, to better understand how technologies work together. IT infrastructure forms the foundational framework of every Information Technology system, encompassing a wide array of physical and virtual components necessary for the functioning of digital services. At the core of this infrastructure are physical hardware elements such as servers, which store and process data, and network devices like routers and switches, which direct and manage the flow of data across networks. Complementing these are computers and other end-user devices that interact with the network. Additionally, physical data centers house these critical hardware components, providing the necessary power, cooling, and security to ensure their optimal operation.

On the virtual side, software plays a pivotal role in the infrastructure. This includes operating systems, which act as the primary interface between hardware and software applications, and various types of software applications that perform specific tasks. Virtualization technologies are also key, allowing for the creation of virtual machines and environments that can efficiently utilize physical hardware resources. Cloud computing has further expanded the boundaries of IT infrastructure by hosting these virtual resources on remote servers, accessible over the internet, offering scalability and flexibility previously unattainable with traditional on-premises setups.

Network architecture, a massive part of the infrastructure, dictates how devices are interconnected and communicate within an organization and over the internet. This includes the design and layout of networks, the protocols used for communication (like TCP/IP), and security measures such as firewalls and intrusion detection systems to protect data. Understanding network architecture is necessary for ensuring that IT systems are not only efficient and reliable but also secure from various digital threats. The basics of IT infrastructure lay the groundwork for more advanced topics in the field, providing a comprehensive understanding of how various technologies and components come together to create efficient systems.

Cloud Computing

Cloud computing has transcended beyond its initial models of data storage and processing, evolving into a dynamic platform offering a wide range of services from infrastructure to software solutions. Today's cloud computing encompasses advanced technologies like artificial intelligence and machine learning, providing unprecedented scalability and flexibility. This evolution has been marked by a shift towards more secure, robust, and user-centric services, catering to a diverse array of business needs. The journey from simple storage services to complex, integrated cloud ecosystems reflect the rapid advancements in this technology, making it a cornerstone of modern IT infrastructure.

The trend in cloud computing has shifted towards more complex architectures like hybrid and multi-cloud strategies. Hybrid cloud combines the flexibility of public cloud with the control of private cloud, offering businesses a balanced approach to their cloud computing needs. Multi-cloud strategies involve using services from multiple cloud providers, maximizing the benefits of each. Meanwhile, edge computing is emerging as a core component, processing data closer to where it's generated, significantly reducing latency and bandwidth usage. These developments represent a move towards more adaptable, efficient, and tailored cloud computing solutions.

Innovations like serverless computing are revolutionizing how businesses deploy applications, allowing them to focus on code rather than managing servers. Advances in cloud security are also notable, with enhanced protocols and AI-driven security measures becoming standard to protect data against increasingly sophisticated cyber threats. Furthermore, the integration of AI and machine learning into cloud services has opened up new frontiers for data analysis and automation. These advancements not only improve the functionality and efficiency of cloud services but also expand the possibilities of what businesses can achieve with cloud computing.

Cloud computing's versatility is evident in its widespread application across various industries. In healthcare, it facilitates telemedicine and patient data management; in finance, it supports secure transactions and fraud detection; in education, it enables remote learning and resource sharing. Looking ahead, the future of cloud computing is likely to see further integration with emerging technologies like IoT and quantum computing, offering even more sophisticated solutions. Sustainable cloud computing is also on the rise, focusing on energy efficiency and reduced environmental impact.

As cloud computing continues to grow, so do the career opportunities in this field. Roles like cloud architect, cloud engineer, and cloud security specialist are in high demand. Skills for these roles include proficiency in cloud service platforms, understanding of cloud architecture, and knowledge of security and compliance. The increasing complexity of cloud solutions also means that skills in automation, AI, and data analytics are becoming increasingly valuable.

Virtual and Augmented Reality

Virtual Reality (VR) and Augmented Reality (AR) are transformative technologies, each offering unique experiences. VR immerses users in a fully digital environment, often using headsets, while AR overlays digital information onto the physical world, typically through smartphone screens or AR glasses. Both technologies have evolved rapidly, driven by advancements in graphics, processing power, and sensor technologies. Understanding the distinction and capabilities of each is important for individuals looking to enter these fields. The immersive nature of VR and the interactive overlay of AR have opened new frontiers in user experience and interaction.

Technological Advancements: The technological landscape of VR and AR is advancing rapidly. Recent developments include more sophisticated headsets with higher resolutions and wider field of view, improved motion tracking, and more realistic haptic feedback. On the software side, advancements in 3D modeling, real-time rendering, and user interface design are enhancing the immersive aspect and usability of VR/AR experiences. These advancements are not only improving the user experience but also expanding the practical applications in various sectors.

Industry Applications: VR and AR have found applications across numerous industries, revolutionizing traditional practices. In gaming, VR offers an immersive experience, while AR games blend digital elements with the real world. In education, these technologies often provide interactive and engaging learning experiences, from virtual field trips to complex scientific simulations. Healthcare uses VR for therapy and surgery simulation, and AR for patient education and surgical assistance. Real estate and interior design are utilizing VR/AR for virtual property tours and space planning, demonstrating the versatility of these technologies.

Challenges and Limitations: While these technologies hold immense promise, they also face challenges such as hardware constraints, high costs, and user accessibility issues. There are ongoing concerns about health effects like motion sickness in VR and privacy issues in AR. Addressing these challenges involves continuous technological refinement and the development of standards and best practices for their usage.

Career Opportunities: The expanding markets are creating diverse career opportunities. Roles include VR/AR software developers, content creators, UX/UI designers, and project managers. These careers require not only technical skills in programming and design but also an understanding of the unique aspects of immersive technologies. As the industry grows, the demand for experts who can bridge the gap between traditional computing and immersive experiences is increasing.

Future Trends and Potential: The future of VR and AR is poised for significant growth, with advancements expected in fields like immersive education, virtual tourism, and augmented retail. As hardware becomes more affordable and user-friendly, these technologies are likely to become more integrated into everyday life. Anticipated developments include improvements in haptic feedback, eye-tracking, and integration with other emerging technologies like AI and 5G.

Ethical and Social Considerations: As these technologies become more prevalent, ethical and social considerations come to the forefront. This includes concerns about digital addiction, the impact on social behavior, and the potential for creating realistic but false representations of people and environments. It's advised that individuals in this field consider the broader implications of their work and strive for responsible development and use of VR and AR technologies.

Artificial Intelligence and Machine Learning

Artificial Intelligence (AI) and Machine Learning (ML) are reshaping numerous industries, offering exciting career opportunities in IT. AI, the broader concept, refers to creating intelligent machines capable of performing tasks that typically require human intelligence. Machine Learning, a subset of AI, involves the development of algorithms that can learn and make predictions or decisions based on data. These technologies are used in diverse applications, from automating mundane tasks to driving innovations in fields like healthcare, finance, and autonomous vehicles.

The career paths in these technologies are varied and require a blend of technical and analytical skills. Fundamental programming knowledge in languages like Python, along with an understanding of data science and statistics, is recommended. However, it's not just about the technical skills; professionals in this field also need to consider ethical implications and the societal impact of their work. For beginners interested in either, starting with basic programming and gradually working into data science concepts is a practical approach. Continuous learning and staying abreast of the latest trends and research are necessary in this rapidly evolving field. As AI and ML continue to grow, they should create new job roles and transform existing ones, making them a core area of focus for anyone looking to build a career in IT. Embracing these technologies can lead to roles like AI developer, machine learning engineer, data analyst, or even specialized positions in AI ethics and policy.

One prevalent concern surrounding AI is its potential impact on employment, with fears that automation and intelligent systems might replace human jobs. However, the reality is more nuanced. AI is creating new job opportunities and augmenting existing roles, making them more efficient. Those who adapt and learn to work alongside it are positioned to thrive. AI doesn't just take over tasks; it opens up new possibilities for innovation and efficiency in various sectors.

AI technologies like chatbots and virtual assistants, have broad applications beyond just the IT industry. They can function as study buddies, aiding in education, as workout or meal planners in personal health, and as research assistants in various fields. These various applications demonstrate the technology's potential to enhance daily life and work, offering personalized assistance and insights. View AI as a tool that complements human skills and creativity, rather than as a replacement.

Emerging Trends

Transitioning into the realm of emerging technologies, we'll embark on a journey through cutting-edge innovations shaping the future of IT. From blockchain technology to the Internet of Things (IoT) and the transformative potential of big data and 5G technology, this section explores the forefront of technological evolution and its profound impact on industries and societies worldwide.

Blockchain Technology

Blockchain technology, often associated primarily with crypto-currencies like Bitcoin, has far-reaching implications well beyond digital currencies. At its core, blockchain is a decentralized ledger technology (DLT) that ensures transparency, security, and integrity of data without the need for a central authority. This makes it a potent tool for various applications in fields such as finance, supply chain management, healthcare, and more.

In finance, blockchain is revolutionizing how transactions are conducted, offering secure, transparent, and efficient methods for transferring assets. Smart contracts, self-executing contracts with terms directly written into code, are a significant innovation in this space, automating contractual processes and reducing the need for intermediaries. Supply chain management is another area where blockchain shows immense promise. It can be used to track the production, shipment, and receipt of products in a transparent and unalterable manner. This improves traceability, reduces fraud, and enhances the overall efficiency of supply chains.

In the realm of healthcare, blockchain can securely manage patient records, ensuring data integrity and easy access while maintaining privacy. It also holds potential for streamlining medical research by providing secure and anonymous data sharing. Also, blockchain's implication in identity management offers a solution to digital identity challenges, allowing for secure and irrefutable identity verification processes. This can greatly enhance security measures in various sectors, from banking to government services.

Blockchain's vast versatility makes it a significant emerging technology in the IT sector, with a growing demand for experts skilled in blockchain development, implementation, and management. Understanding blockchain's potential and its applicability across different industries is beneficial for anyone looking to be at the forefront of technological innovation in IT.

A Real-World Example: Facing challenges with product authenticity and tracking, a company adopted a blockchain-based system to ensure transparency and traceability. This system allowed for real-time tracking of products from manufacturer to store shelves, drastically reducing instances of counterfeit goods. The blockchain's immutable ledger provided a reliable and unalterable record of each product's journey, enhancing trust among suppliers, retailers, and customers. This implementation not only streamlined the business' supply chain operations but also significantly improved inventory management and customer satisfaction. The success of this project illustrates the practical application of blockchain technology beyond cryptocurrencies, showcasing its potential to solve real world business challenges.

Internet of Things (IoT)

The Internet of Things (IoT) refers to the network of physical devices connected to the internet, capable of collecting and sharing data. This technology encompasses a wide range of devices, from everyday household items to sophisticated industrial tools. IoT devices are embedded with sensors, software, and other technologies to connect and exchange data with other devices and systems over the internet. The concept of IoT extends connectivity beyond traditional devices like computers and smartphones, integrating with the physical environment in innovative ways.

Everyday Applications: IoT has seamlessly integrated into daily life, revolutionizing how we interact with our environment. Smart home devices like thermostats, lights, and security systems can be controlled remotely and automated for enhanced convenience and energy efficiency. Wearables such as fitness trackers monitor health metrics and sync with mobile apps to provide personalized health insights. Even everyday appliances like refrigerators and washing machines are becoming smarter, with the ability to communicate usage patterns, automate tasks, and provide alerts for maintenance.

Impact on Industries: The influence of IoT extends far beyond personal convenience, significantly impacting various industries. In agriculture, IoT devices can monitor soil moisture, weather conditions, and crop growth, enabling precision farming for increased efficiency and yield. Healthcare has seen advancements with remote patient monitoring and real-time data analysis, improving patient care and outcomes. Manufacturing industries utilize IoT for predictive maintenance of machinery, real-time tracking of inventory, and optimizing production processes. This widespread adoption in different sectors is enhancing operational efficiencies, reducing costs, and opening new avenues for innovation.

Data Collection and Analysis: IoT devices are integral to big data, constantly collecting a wealth of information from their environments. This data, when analyzed, can reveal patterns, trends, and insights that were previously inaccessible. In a smart home these devices can track energy usage over time, enabling homeowners to optimize consumption and reduce costs. In industrial settings, data collected from IoT sensors can be used to improve operational efficiency, predict equipment failures, and streamline supply chains. The ability to analyze and act upon the data collected these devices is transforming decision-making processes in both business and everyday life.

Consumer Benefits: IoT offers numerous benefits to consumers, enhancing convenience, safety, and efficiency. Smart home devices, for instance, allow users to automate and remotely control home systems, leading to increased convenience and energy savings. Wearable technology provides personalized health monitoring, helping users make informed decisions about their lifestyle. In terms of safety, devices like smart smoke detectors and security systems offer real-time monitoring and alerts, providing peace of mind.

Smart Cities: IoT technology is a pillar in the development of smart cities. It enables more efficient use of resources, improved traffic management, enhanced public safety, and better environmental monitoring. IoT sensors can monitor traffic flow and optimize street lighting and energy use, reducing the city's carbon footprint. Public safety is enhanced through connected cameras and emergency response systems that can react faster and more efficiently. Additionally, these devices play a huge role in monitoring environmental conditions, helping cities respond to issues like air quality and water management proactively.

Security Concerns: With the proliferation of smart devices, security has become a paramount concern. These devices, if not properly secured, can be vulnerable to hacking, leading to privacy breaches and data theft. For instance, unsecured smart home devices can be entry points for cybercriminals into personal networks. In industrial settings, a compromised smart device could lead to significant operational disruptions. Addressing these security concerns involves implementing robust security protocols, regular software updates, and educating consumers about safe IoT practices.

Future Trends: The future of IoT holds immense potential, with advancements like edge computing, which processes data closer to where it is collected, reducing latency. Integration with AI will further enhance smart device capabilities, enabling more intelligent and autonomous decision-making by devices. Another emerging trend is the development of IoT standards and frameworks to ensure interoperability between devices and systems.

Career Opportunities: IoT's growth is creating a wide array of career opportunities. These include roles in device design and development, network architecture specialized for smart devices, data analytics focusing on IoT-generated data, and cybersecurity tailored to protect IoT networks. Professionals with skills in programming, network engineering, data science, and security are increasingly in demand in this domain.

Challenges and Considerations: Despite its benefits, IoT implementation comes with challenges such as interoperability issues between different devices and platforms, managing the vast amount of data generated, and ensuring user privacy and data security. There's also a need for standardized regulations in development and usage. Addressing these challenges is critical for the successful integration of IoT technology into daily life and business operations.

Big Data and Analytics

Big data refers to extremely large data sets that traditional data processing software cannot handle effectively. It's characterized by the four Vs: volume (immense scale), variety (different forms of data), velocity (speed of generation and processing), and veracity (uncertainty of data). Big data's complexity requires specialized tools and technologies for processing and analysis. Understanding big data can help provide insights that can lead to better decision-making and strategic business moves. The ability to manage and analyze big data is becoming increasingly important across various sectors.

Technologies in Big Data: Technologies like Hadoop and Spark are instrumental in handling big data. Hadoop, an open-source framework, allows for distributed storage and processing of large data sets across computer clusters using simple programming models. Spark, another open-source distributed computing system, provides an interface for programming entire clusters with implicit data parallelism and fault tolerance. Both are designed to scale up from single servers to thousands of machines. Specialists should have familiarity with these technologies when working with big data, as they provide the foundation for managing and analyzing large data sets.

Data Analytics: This involves examining large data sets to uncover hidden patterns, correlations, and other insights. With today's technology, it's possible to analyze your data and get answers almost immediately – an effort that's slower and less efficient with more traditional business intelligence solutions. This instantaneous data processing capability enables businesses to make informed decisions more quickly. In a world increasingly driven by data, analytics plays a lead role in understanding and leveraging information for strategic advantage.

Big Data in Various Sectors: In healthcare, big data enables the analysis of patient data to improve treatment and predict outbreaks. In retail, it helps in understanding customer behavior and preferences. Financial institutions use big data for risk management and fraud detection. Additionally, in sectors like telecommunications and manufacturing, big data is used for optimizing operations and predictive maintenance. The application of big data in these sectors not only enhances efficiency but also drives innovation and growth.

Challenges & Ethical Considerations: Managing big data presents many challenges, especially in terms of privacy, security, and data integrity. Ethical considerations are paramount, as the misuse of data can lead to violations of privacy and biased decision-making. Ensuring the accuracy and ethical use of big data, while safeguarding user privacy, is a growing concern. There's also the challenge of sifting through vast amounts of data to find relevant and accurate information, requiring sophisticated algorithms and skilled professionals.

Future Trends: The future of big data is likely to be influenced by advancements in AI and machine learning, enhancing the ability to analyze complex data sets. The integration of IoT with big data will also continue to grow, providing even more data sources. Another trend is the increasing importance of real-time data processing for immediate decision-making. These trends suggest a future where big data becomes even more integral to business strategies and operations.

Skills & Careers: This field offers a range of career opportunities, including data scientists, big data engineers, and data analysts. Core skills include proficiency in data processing frameworks like Hadoop and Spark, expertise in statistical analysis, and the ability to derive insights from large data sets. As the field grows, there's an increasing demand for specialists who can not only handle the technical aspects of but also understand its business applications and ethical implications.

5G Technology

5G is more than the next step in mobile network technology; it's a fundamental shift in wireless communication. Designed for faster speeds, higher capacity, and lower latency, 5G is set to revolutionize how we interact with technology. Unlike previous generations, this technology aims to provide a versatile infrastructure that supports a wide range of applications, from enhanced mobile broadband to massive IoT deployments and mission-critical communications. This new network ecosystem is expected to be a key driver in the next wave of digital transformation, impacting industries and consumers alike.

Advancements and Features: 5G's advancements include unprecedented speed, enabling users to download content faster than ever. Its low latency enhances responsiveness in applications requiring real-time feedback, crucial for technologies like autonomous vehicles. Additionally, its capacity to support a vast number of connected devices simultaneously is important for expanding smart networks. These features collectively facilitate new applications and services, from augmented reality experiences to smart city infrastructures, setting the stage for a more connected future.

Impact on IoT and Smart Devices: The rollout of 5G is expected to significantly accelerate the growth and capabilities of IoT. With its high speed and low latency, it enables more efficient and reliable communication between devices. This enhanced connectivity fills a need for applications requiring real-time data exchange, such as autonomous vehicles and industrial automation. The proliferation of smart devices, powered by 5G, will lead to smarter homes, cities, and industries, where automation and data-driven decision-making become the norm.

Enhanced Mobile Broadband: 5G's high data rates will enable smooth streaming of high-definition videos, and its low latency will greatly benefit online gaming and virtual reality applications. The enhanced capacity of these networks also means that congested areas, like stadiums and concert venues, will see improved connectivity. This technological leap in mobile broadband is not just about speed; it's about creating new opportunities for content consumption and digital interaction.

Challenges and Deployment Issues: Deploying 5G involves overcoming many challenges. The infrastructure upgrade to support 5G is extensive and costly, requiring new cell towers and updated network equipment. Additionally, the need for closer cell tower spacing due to its higher frequency bands pose logistical challenges, particularly in dense urban areas. There are also concerns about the availability and allocation of radio frequency spectrum, which is critical for 5G's operation. Overcoming these hurdles is key to realizing the full potential of this technology.

Potential Applications & Use Cases: 5G's potential extends across a variety of applications. In healthcare, it could enable telemedicine and remote patient monitoring, making healthcare more accessible. In the automotive industry, it is needed for the development of connected and autonomous vehicles, enhancing road safety and traffic efficiency. In entertainment, it opens new possibilities for immersive experiences with AR and VR.

Future Outlook: The future of 5G technology promises to bring about significant changes in the way we interact with the digital world. As 5G networks become more widespread, they will lay the groundwork for smart cities, where everything from traffic lights to utility services will be interconnected. The integration with emerging technologies like AI and IoT will lead to more intelligent and autonomous systems. This evolution will not only enhance existing applications but also pave the way for innovations yet to be imagined.

Career Opportunities: The expansion of this technology is creating new job opportunities and requiring new skills. Professionals with expertise in network engineering, 5G infrastructure deployment, and wireless communication are in high demand. There is also a growing need for skills in developing 5G-enabled applications and services. As technology evolves, staying informed of the latest developments and acquiring relevant skills will be valuable for those seeking to make a mark in this rapidly advancing field.

Integration and Interoperability

In the landscape of Information Technology, understanding how different technologies integrate and work together is as important as understanding the technologies themselves. This section on integration and interoperability is valuable because it highlights how the convergence of technologies like AI, IoT, Cloud Computing, and others is creating more complex and efficient systems, a trend that is reshaping the industry.

Technological integration is about creating synergies where the sum is greater than its parts. It involves combining technologies like AI, IoT, and Cloud Computing to enhance functionality and efficiency. For instance, integrating AI with IoT devices leads to smarter, more responsive systems. This convergence is powering advancements in areas ranging from smart homes to industrial automation, demonstrating the transformative potential of integrated technologies.

Interoperability Trials: Addressing the interoperability challenges in IT can require a multifaceted approach. Firstly, it involves developing and adhering to universal standards and protocols, which facilitate communication between different technologies and systems. Industry collaboration is key, with organizations and standard-setting bodies working together to establish these common guidelines. Secondly, adopting open-source technologies can also aid in enhancing interoperability, as they allow for greater flexibility and customization. The importance of resolving these challenges lies in the ability to create efficient, cohesive systems that maximize the benefits of technological integration. It enables seamless data exchange and system functionality, critical in applications ranging from healthcare to smart city infrastructures.

Standardization Efforts: Developing common standards and protocols ensures that different technologies can communicate and function together effectively. This is particularly important in areas like IoT, where a myriad of devices and platforms need to interact. Standardization efforts by international bodies and industry groups help in streamlining integration across the IT landscape.

Future of Integrated Technologies: Potential advancements could further revolutionize how we interact with technology. As integration deepens, we can expect more autonomous, intelligent systems capable of complex tasks and decision-making. This evolution will likely lead to more innovative applications, driving progress in fields like smart cities, healthcare, and beyond.

In wrapping up this chapter on Current and Emerging Technologies, we've navigated through the complexities and potential of various cutting-edge technologies, understanding not just their standalone functions but also how their integration shapes the IT industry. As we proceed to Chapter 3, the emphasis will shift to acquiring the necessary knowledge and skills. This will prepare you to engage effectively with these technologies, offering insights into educational paths that align with the evolving demands of the industry. It's an exciting journey ahead, where learning and adaptation will help you to thrive in this dynamic digital scene.

Reflect and Apply Questions

1. **Exploring AI and ML:** Consider a recent encounter with AI in your daily life, such as a virtual assistant or recommendation algorithm. How do you see these technologies impacting future careers in IT and other industries?
2. **Blockchain Application:** Reflect on a transaction or process in your life (like banking or online shopping). How do you think blockchain technology could improve the security and efficiency of such processes?
3. **IoT in Everyday Life:** Identify an IoT device you use or are familiar with. How does this device make your life easier, and what potential do you see for smart devices to transform other aspects of daily living?
4. **Integration and Interoperability:** Consider a piece of technology you use that integrates with other devices or platforms. How does this integration enhance its usefulness, and what challenges do you think exist in ensuring different technologies work seamlessly together?

Chapter 3: Building the Right Educational Base

This chapter explores the steps in constructing a solid educational foundation for a successful career in Information Technology. We will begin by discussing various academic programs options, from degrees to certifications, and how to choose the right one for your goals. We'll emphasize the importance of creating a personalized learning roadmap tailored to your career aspirations. Seeking mentorship from professors and experts, contributing to open-source projects, and strategically pursuing technology certifications are also core topics in this chapter. By the end, you'll have a clear understanding of how to shape your educational journey to thrive in the IT field.

Exploring Educational Opportunities

This section covers various paths available for individuals seeking to build a strong educational foundation in IT. It encompasses academic options, certifications, and practical learning experiences through hands-on projects and labs. Additionally, readers will explore personalized learning strategies and the role of mentorship from seasoned professionals.

Academic Options and Specializations

Bachelor's degrees in information technology typically span four years and offer a comprehensive education in various disciplines. These programs provide students with a broad foundation in computer science, mathematics, programming, and IT management. They often include general education courses, which offer a well-rounded education beyond the tech field. Bachelor's degrees are an excellent choice for individuals looking for a deep and holistic understanding of IT concepts, which can be advantageous for roles that require a wide range of skills, such as management or systems analysis.

Associate degree programs in are shorter in duration, typically taking two years to complete. They offer a more focused education in IT subjects and are ideal for individuals looking to enter the workforce relatively quickly or as a stepping stone to further education. These programs cover core topics and often include practical, hands-on experience. Associate degrees are a cost-effective way to gain foundational knowledge and can lead to entry-level roles in IT support, helpdesk, or network administration.

Degree Program Real World Example: Ally, a recent computer science graduate, leveraged her university's diverse curriculum to gain a broad understanding of IT. She actively participated in research projects, which honed her technical skills and led to a groundbreaking senior thesis in machine learning. Her disciplined approach to note-taking and research helped her excel academically and secure a role at a leading tech firm.

Certificates: IT certifications are specialized, short-term programs that concentrate on specific skills or technologies. They are designed to provide targeted education in a shorter time frame than degree programs. These certificates are highly focused and can be an excellent choice for individuals who want to acquire specific skills quickly. These certifications can enhance job prospects and are often used to demonstrate proficiency in a particular area of IT. More on this later in this chapter.

Certification Journey Real World Example: Bill, with a keen interest in IT but no formal education in the field, embarked on a journey to gain industry recognition through certifications. Starting with the CompTIA A+ to lay a strong foundation, he then ventured into the world of cybersecurity. Balancing a full-time job in a tech support role, Bill dedicated his evenings and weekends to studying. He actively participated in online forums for deeper understanding. As he progressed, he achieved more advanced certifications like Certified Ethical Hacker (CEH) and CISSP. These certifications not only validated his self-taught skills but also opened doors to higher level positions. Bill's story exemplifies how dedication to continuous learning can significantly elevate a career in the industry.

Bootcamps: These are intensive, short-duration programs that focus on teaching practical, job-specific skills. They are known for their immersive and accelerated nature, making them a popular choice for individuals seeking rapid career entry or transitions. IT bootcamps cover a range of topics, including coding, web development, data science, and cybersecurity. These programs often include hands-on projects and emphasize real-world application. While bootcamps offer quick skill acquisition, they may not provide the depth of knowledge found in traditional degree programs, making them a valuable option for those looking to start their careers quickly.

In your journey through various educational pathways, whether it's a degree, certification, or bootcamp, the significance of individual research and diligent note-taking cannot be overstated. These practices are invaluable to navigating a vast ocean of information as they provide direction and ensure you retain and comprehend the wealth of knowledge you encounter. Effective note-taking helps in organizing complex concepts, preparing for exams, and tackling real-world IT problems. Meanwhile, active research deepens your understanding, keeps you updated with the latest trends, and fosters an inquisitive mindset, crucial for success in the ever-evolving sector. Developing these skills early in your educational journey will enhance your learning experience and lay a strong foundation for your future career.

Let's dive deeper into some of the IT specializations frequently found when pursuing a degree. Keep in mind, these can vary greatly between College's/University's. Some have minimal options, while others have a vast selection to choose from.

Cybersecurity: A specialization in cybersecurity is dedicated to safeguarding digital assets and data from cyber threats. It covers a wide range of topics, including network security, ethical hacking, risk assessment, and digital forensics. Students learn how to identify vulnerabilities in systems, protect against malicious attacks, and respond to security incidents. With the increasing importance of data protection and privacy, cybersecurity specialists are in high demand across various industries, making it a promising career path.

Software Development: This is a dynamic field that specializes in creating applications and robust software solutions. This specialization involves learning programming languages, software design principles, and development methodologies. Students may choose to focus on specific areas such as mobile app development, web development, game development, or even software engineering. As technology continues to advance, software developers play an important role in shaping the digital landscape by creating innovative and user-friendly software products.

Data Science: This field is all about extracting valuable insights from vast datasets. This specialization covers statistical analysis, machine learning, data visualization, and data engineering. Data scientists are skilled in using tools and techniques to analyze data, make predictions, and support data-driven decision making. With the increasing availability of data in today's world, data science specialists are in high demand across industries such as healthcare, finance, and ecommerce.

Networking and Infrastructure: Specialists in this field are responsible for designing, implementing, and managing the technical backbone of IT systems. This specialization explores topics like network architecture, server administration, cloud computing, and virtualization. Students learn how to build and maintain secure and efficient computer networks, ensuring that data and information flow seamlessly. As organizations rely more on digital infrastructure, professionals in this field play a huge role in keeping systems operational and secure.

IT Project Management: This trade is all about efficiently planning, executing, and delivering IT projects. This specialization focuses on project management methodologies, resource allocation, risk assessment, and stakeholder communication. IT project managers ensure that projects are completed on time, within budget, and meet the desired objectives. They can specialize in various types of projects, including software development, infra-structure deployment, or data analytics initiatives. With IT projects becoming increasingly complex, skilled project managers are vital for successful project outcomes.

Cloud Computing: Students specializing in cloud computing learn how to design, deploy, and manage cloud infrastructure and services. They gain expertise in virtualization, containerization, and cloud security. Additionally, cloud experts develop skills in optimizing cloud resources, ensuring high availability, and managing cloud-based applications and data.

These concentrations offer students the opportunity to tailor their education and skills to align with their career aspirations in the diverse field of Information Technology. Each specialization brings its unique challenges and opportunities for those looking to make an impact in the digital world.

There is an increasing trend of curriculums that include industry relevant certifications. Combining these within learning material makes for a huge benefit for students looking to get into the IT field. Certifications from reputable organizations (more on that below) provide a leg up on the competition that may only have a degree, or no degree at all. When researching course outlines, it would be beneficial to find which paths include certifications. Then research those certifications to make sure that material is in line with the career path you are interested in. Often times, there will be various concentrations with different certifications included. This can also be a cost saver as you won't have to pay for the certifications separately.

Certifications and Their Relevance

In today's competitive job market, IT certifications have gained substantial importance as a means of validating an individual's technical skills and knowledge. These certifications serve as tangible proof of one's expertise in a specific domain. They are recognized by employers worldwide and are often used as a benchmark to assess a candidate's qualifications during the hiring process. These certifications offer several advantages, including increased employability, enhanced earning potential, and career advancement opportunities. Employers value certified professionals as they bring a level of credibility and competence to their organizations.

Moreover, IT certifications are particularly valuable because they provide a structured learning path. This designed approach allows students to acquire specialized knowledge and skills in a systematic manner. It not only helps them become proficient in their chosen field but also keeps them updated with the latest industry trends and technologies. Whether you're aiming to become a network administrator, cybersecurity expert, cloud architect, or a developer, pursuing relevant certifications is a clear pathway to success in the industry.

Types of Certifications: Depending on your career goals and interests, you can choose from a wide range of certification paths. Some common categories of IT certifications include:

- **Networking:** These certifications focus on networking technologies, protocols, and infrastructure. Examples include CompTIA Network+, Cisco CCNA, and Juniper JNCIA.
- **Security:** These are invaluable in today's cyber-threat landscape. Certifications like CompTIA's Security+, ISC2's Certified Information Systems Security Professional (CISSP), and the Certified Ethical Hacker (CEH) validate skills in protecting systems and data.
- **Cloud:** With the increasing adoption of cloud computing, certifications in platforms like AWS, Microsoft Azure, and Google Cloud are highly sought after.
- **Programming and Development:** Developers can pursue certifications in programming languages (e.g., Java, Python), web development (e.g., CompTIA Web

Developer), or mobile app development (e.g., AWS Certified Mobile Development).

- **Database:** For those interested in database management, certifications like Oracle Certified Associate (OCA) or Microsoft Certified: Azure Database Administrator Associate are available.
- **Project Management:** IT project managers can benefit from certifications such as Project Management Professional (PMP) or Certified ScrumMaster (CSM).

These are just a few examples, and there are certifications tailored to nearly every IT specialization imaginable. Choosing the right certification path is an important step in aligning your career with your interests and goals.

Recognized Certification Bodies: When pursuing certifications, it's important to choose from reputable organizations or certification bodies. These organizations have established themselves as industry leaders in providing high-quality certification programs. Recognized bodies ensure that their accreditations adhere to rigorous standards and accurately reflect the skills and knowledge required in the field. Some well-known bodies in the IT industry include:

- **CompTIA:** Known for certifications like A+, Network+, and Security+, CompTIA offers entry-level to advanced certs covering a wide range of IT topics.
- **Cisco:** Cisco certifications, such as CCNA and CCNP, are highly regarded in the networking field. Cisco also has many other specializations like Security.
- **Microsoft:** Microsoft certs validate skills in Microsoft technologies, including Azure, Windows, and Office 365.
- **AWS (Amazon Web Services):** AWS certifications certify expertise in cloud computing and AWS services.
- **(ISC)²:** This organization offers highly regarded cybersecurity certifications like the CISSP and Certified Cloud Security Professional (CCSP).
- **ISACA:** ISACA certs, including CISA and CISM, focus on information security, governance, and risk management.

Pursuing accreditations from these reputable bodies will ensure that your efforts are recognized and respected within the industry, increasing your credibility and employability.

Certification Paths: The journey through IT utilizing certifications is often a structured progression, allowing professionals to build a comprehensive skill set. Starting with foundational certs like CompTIA's A+ for general IT knowledge, individuals can then specialize, perhaps advancing to Cisco's CCNA for networking or AWS's Certified Solutions Architect in cloud computing. This systematic approach not only equips professionals with relevant skills but also demonstrates a commitment to career growth and adaptability in the ever-changing IT field. There is a vast number of resources available in regards to certification paths. It is encouraged to do research, watch videos, and talk with experts in the industry. One benefit of accreditations is you can take many from different paths to broaden your knowledgebase.

Below are some mock certification paths. Keep in mind everyone's path is different; some find it best to start with fundamentals while some who already have knowledge start with higher-level certs.

1. **Cybersecurity Path:**
    - Start with CompTIA Security+ for foundational security concepts.
    - Progress to Certified Ethical Hacker (CEH) for skills in penetration testing and ethical hacking.
    - Obtain Certified Information Systems Security Professional (CISSP) for a deeper understanding of overall security management.
    - Consider specialized certifications like Certified Cloud Security Professional (CCSP) if interested in cloud security aspects.
2. **Networking Path:**
    - Start with CompTIA's A+ for foundational IT skills
    - Advance to Cisco's CCNA to gain core networking skills.
    - Advance to Cisco Certified Network Professional (CCNP) for more in-depth knowledge.
    - Explore further with certifications like CompTIA Network+ or Juniper Networks Certified Internet Associate (JNCIA) for broader networking concepts.
3. **Cloud Computing Path:**
    - Start with AWS Certified Clout Practitioner for fundamental cloud concepts.
    - Next, take AWS Certified Solutions Architect Associate for more advanced topics.

- Proceed to AWS Certified Solutions Architect - Professional for expert cloud architecture knowledge.
- Expand skills with Microsoft Certified: Azure Administrator Associate for cross-platform expertise.

4. **Data Science Path:**
   - Initiate with Microsoft Certified: Azure Data Scientist Associate for foundational data science skills.
   - Progress to certifications like SAS Certified Data Scientist for advanced analytics techniques.
   - Consider IBM Data Science Professional Certificate to diversify your skill set in the data science domain.

A Certification Journey: Mia began her IT career in a help desk role. Eager to advance, Mia pursued the CompTIA Network+ certification to gain foundational networking knowledge. After achieving this certification, she was able to move into a junior network support role, where she gained practical experience. Motivated by her progress, Mia then tackled the more advanced Cisco Certified Network Associate (CCNA) certification. This accreditation deepened her understanding of networking concepts and practices, leading to a promotion to a network engineer position within her company. Mia's story demonstrates how targeted IT certs can significantly impact career advancement. Her journey from a beginner in IT to a skilled network engineer underscores the value of certifications in providing career growth opportunities and enhancing job prospects in the IT industry.

My Story: I had 5 years of experience before I obtained my first certification, Network+. I then proceeded to get the CCNA, expecting to continue my career in networking. After no luck progressing in that field, I jumped into cybersecurity. Within a year I obtained Security+, CySA+, CSAP, PCNSA, and the CISSP. My work ethic and commitment to learning earned me a position at a prominent cybersecurity company. The CISSP is the gold standard in cyber, but being committed to continuous learning I have my sites set on the CISM and Microsoft Azure certifications to broaden my knowledgebase.

Benefits of Accreditation: Earning IT certs can lead to substantial career benefits. They open doors to new job opportunities, often with higher salaries, as they validate skills and knowledge to employers. Certifications can also lead to promotions and greater job security. Beyond financial gains, they also contribute to professional growth, keeping individuals updated with the latest technologies and practices in their field.

Certification Preparation: Utilizing study guides, online courses, and practice exams is the common approach. Many opt for structured training programs offered by certification bodies or educational institutions. Joining study groups and participating in forums can provide additional insights and support. The key is a consistent and focused study plan tailored to the specific certification.

Continuous Learning: Certifications require regular renewal, often through continued education or re-examination, ensuring professionals stay current with technological advancements. This commitment to ongoing learning fosters adaptability and resilience, core traits for thriving in the industry.

Certification vs. Degree: While both certifications and degrees have their merits, they serve different purposes. A degree provides a broad, theoretical foundation and is often favored for entry-level positions and certain career paths. Certifications, on the other hand, are more focused and practical, ideal for specific skill enhancement and staying current in the field. They can complement a degree or serve as a standalone testament to one's expertise in niche areas.

Practical Learning Through Hands-On Projects and Labs

Hands-on experience in IT is invaluable, as it bridges the gap between theoretical knowledge and real-world application. By engaging in practical projects and labs, learners gain a tangible understanding of core concepts, which is often missed in purely academic settings. This approach allows for a deeper comprehension of complex subjects, as learners can see the direct impact of their actions in a controlled environment. It also prepares them for the challenges they will face in professional roles, as they learn to apply their knowledge in practical, often unpredictable, situations. Furthermore, hands-on experience enhances problem-solving skills, as learners must navigate and resolve real-world issues.

Project-Based Learning: To find and start relevant projects, learners can begin with simple tasks that match their current level of understanding. Online platforms like GitHub offer a plethora of open-source projects that vary in complexity and area of focus. Beginners can start with small, manageable projects such as creating a basic website or a simple database and gradually progress to more complex tasks. Local community groups or online forums can be a valuable resource for finding project ideas and collaborators. It's important to choose projects that align with one's interests and career goals, as this increases engagement and the likelihood of completion.

Lab Environments: Setting up a home lab is an excellent way for learners to gain hands-on experience. To start, one doesn't need a sophisticated setup; a basic computer with adequate processing power and memory can suffice. Virtualization software, like VMware or VirtualBox, allows learners to create and manage multiple virtual machines, simulating different operating systems and network environments. Cloud-based services, such as AWS or Azure, offer free tiers or limited-time trials which are excellent for practicing cloud computing concepts. Online resources, forums, and communities can provide guidance on setting up and troubleshooting home labs. The key is to start small and expand the lab as skills and understanding grow.

Integration with Coursework: For a comprehensive learning experience consider integrating hands-on projects with coursework. This integration helps in solidifying theoretical knowledge through practical application. Educators can design coursework that includes real-world projects, encouraging students to apply what they've learned in class to solve practical problems. For instance, a course on networking could include a project where students set up and manage a small network. This approach enhances understanding and prepares students for the type of work they will encounter in their professional lives. It's important for coursework to be aligned with industry trends and technologies, ensuring that the skills and knowledge acquired are relevant and up-to-date.

Examples of Practical Projects: Hands-on projects can range from simple to complex, tailored to different skill levels and interests. For beginners, projects like setting up a personal website or a basic network can provide foundational skills. Intermediate learners might explore building a simple application or configuring a small-scale cloud environment. Advanced projects could involve developing complex software, implementing cyber-security measures, or managing large scale networks. These projects reinforce theoretical knowledge and provide insight into practical challenges and problem-solving in real-world IT scenarios.

Collaborative Projects: These projects mirror the teamwork often required in professional environments. They encourage communication, delegation, and cooperative problem-solving. Working on group projects, such as developing a piece of software or setting up a network system, will help learners understand diverse perspectives and solutions. It also prepares them for the collaborative nature of many IT jobs, where cross functional teamwork and communication are directly related to project success.

Learning from Failure: The educational value of troubleshooting and learning from failures in IT projects cannot be overstated. Hands-on projects often involve unforeseen challenges and issues, offering learners a chance to develop critical problem-solving skills. This process of identifying, addressing, and learning from errors cultivates a resilient and adaptive mindset. It also provides a safe environment for experimentation, where the consequences of failure are minimal but the learning outcomes are significant.

Resources for Hands-On Learning: There are numerous resources available for hands-on learning. Online platforms like Coursera and Udemy offer courses with practical components. Virtual labs and simulators provide realistic environments for practicing skills without the need for physical hardware. Open-source projects on platforms like GitHub present opportunities for real world application and collaboration. Additionally, communities and forums like Stack Overflow and Reddit's IT communities can be invaluable for advice, support, and finding project ideas.

Assessment and Feedback: Regular evaluations, either through peer review or instructor feedback, help learners identify areas for improvement and gauge their progress. Effective feedback provides constructive criticism and guidance, enabling learners to refine their skills and approaches. This process not only enhances technical proficiency but also develops critical thinking and self-assessment skills, which are vital for a successful career in IT.

My Story: When I was preparing for my networking certifications, I was advised a home lab would help tremendously in preparation. I purchased some older Cisco routers, switches, and a server on a marketplace from someone in my city. It was a very educational experience learning how to configure actual equipment network engineers work with. I made many mistakes, but learned so much from the home lab. I have also utilized virtual software to spin up virtual machines on my laptop to gain experience with Linux systems. In my experience, those who take the time to work with a home lab, whether its physical equipment or virtual equipment, will reap the benefits. It is also an excellent talking point in interviews showcasing your commitment to learning proactively and learning from your failures.

Mentorship and Guidance from IT Professionals

Identifying the right mentor in the IT field requires a strategic approach. Look for professionals whose career paths align with your interests and who possess the experience and knowledge you aspire to gain. Networking events, professional associations, and online forums are excellent places to start. Once you've identified potential mentors, it's important to set clear goals for the mentorship. These goals might include learning specific skills, gaining insights into a particular domain, or understanding the nuances of career advancement in IT. A clear set of objectives will guide the mentorship connection, ensuring it is mutually beneficial and focused.

Benefits of Professional Mentorship: Mentors provide insights into industry trends, offer practical advice based on their experiences, and help navigate the complexities of the IT landscape. They can share lessons learned from their own career journeys, including successes and failures, offering a real-world perspective that is invaluable. This mentorship is not just about acquiring technical skills but also about understanding the industry's culture, ethical practices, and professional standards. Learning from someone who has already walked the path can significantly accelerate your development and provide clarity in career decisions.

Effective Communication: This involves not only expressing your goals and challenges clearly but also actively listening to your mentor's advice and feedback. Be open to constructive criticism, as it is crucial for career growth and learning. Regular, honest, and respectful communication helps in building trust and ensures that the relationship is productive. Feedback from an experienced IT expert can provide you with new perspectives, challenge your thinking, and encourage you to step out of your comfort zone.

Mentorship for Career Advancement and Specialized Domains: A mentor can guide you in choosing the right cert-ifications, projects, and learning paths that align with your career goals. They can offer insights into niche areas like cybersecurity, data science, or cloud computing, where specific knowledge and skills are highly valued. Their guidance can help you avoid common pitfalls and make informed decisions about your professional journey. Additionally, mentors can offer invaluable advice on how to stay relevant and competitive in these evolving fields.

Fostering Professional Relationships: Building long term professional relationships with mentors can provide ongoing support and guidance throughout your career. These relationships often evolve beyond the initial learning objectives, offering a lasting network of professional support. A long-term mentor can help you navigate career transitions, provide references, and introduce you to other industry specialists. It's important to nurture these relationships with regular communication, expressing gratitude, and, when possible, giving back to your mentors. Such relationships are priceless assets in the industry, offering both personal and professional enrichment over time.

Harnessing Learning Resources

Transitioning into practical resources, this section focuses on leveraging available tools and platforms to enhance learning. It encompasses contributions to open-source projects and the utilization of online learning platforms and resources. By harnessing these diverse resources, individuals can develop comprehensive skills and knowledge in the field of information technology.

Building a Personalized Learning Roadmap

The first step in creating a learning roadmap is to understand one's personal interests and career objectives. This self-reflection helps in aligning the learning process with individual passions and professional aspirations. Whether you are drawn to cybersecurity, data analysis, or software development, recognizing these interests is beneficial. It guides the choice of specialized areas to focus on and helps maintain motivation throughout the learning journey.

Skills Gap Analysis: This involves evaluating one's current skills against the requirements of the desired role. By identifying areas of weakness or lack of knowledge, learners can create a targeted roadmap focused on acquiring the necessary skills. This process not only streamlines the learning path but also ensures that all efforts are directed towards closing the critical skills gaps.

Researching IT Trends and Demands: IT professionals should always keep up-to-date on the latest trends and market demands. This research guides students in understanding which skills are most valued and in demand. It's important to stay informed about emerging technologies and shifts in the tech landscape, as this knowledge influences the relevance and utility of the learning road-map.

Setting Realistic and Achievable Goals: Setting **SMART** goals is important in a learning roadmap. These goals should be specific, measurable, achievable, relevant, and time-bound.

Specific: Goals should be well-defined and clear to anyone who has a basic understanding of the project or learning path.

- **Non-Specific Goal**: I want to improve my cybersecurity skills.
- **Specific Goal**: I want to become proficient in network security by learning about firewalls and intrusion detection systems.

Measurable: Establish concrete criteria for measuring progress toward the attainment of each goal. This makes it easier to track progress and know when the goal has been achieved.

- **Non-Measurable Goal**: Learn about cybersecurity tools.
- **Measurable Goal**: Achieve accreditation in using Kali Linux and Wireshark for network analysis by completing a certified online course and passing the examination within 6 months.

Achievable: The goal should be realistically attainable given your current resources, knowledge, and time frame. This means setting a goal that is challenging but possible.

- **Unachievable Goal**: Master all cybersecurity domains in six months.
- **Achievable Goal**: Complete an introductory course on cybersecurity fundamentals within three months, dedicating 5 hours a week to study.

Relevant: Ensure the goal matters to you and aligns with other relevant goals. A goal that supports or is in alignment with other objectives is considered relevant.

- **Irrelevant Goal**: "Learning advanced cryptography tech-niques" for someone primarily interested in becoming a network administrator.

- **Relevant Goal**: Understanding basic encryption standards and protocols to secure network communications as a network administrator.

Time-bound: Every goal should have a deadline or a time frame to focus on and something to work toward.

- **Not Time-bound**: I'll learn Python scripting eventually.
- **Time-bound**: I will learn Python scripting within the next four months by completing two programming projects.

Implementing SMART Goals in a Learning Roadmap:

1. **Initial Assessment**: Begin with a self-assessment to identify your current skills and knowledge gaps. This will help tailor your goals to be both relevant and achievable.
2. **Research and Planning**: Investigate potential courses, certifications, and resources that align with your specific goals. Plan how each step fits into your broader career aspirations in IT.
3. **Scheduling and Time Management**: Allocate specific times for study and practice. Use project management tools or calendars to keep track of deadlines and milestones.
4. **Monitoring Progress**: Regularly review your goals and progress. This could be through self-testing, project completion, or feedback from peers or mentors.
5. **Adjustments and Flexibility**: Be prepared to adjust your goals as you progress. Learning is a dynamic process, and staying flexible allows you to adapt to new challenges or interests.

By setting SMART goals, you establish a clear path forward, making the learning process more structured and manageable. This approach not only applies to individual learning endeavors but can also be an effective framework for educators designing curricula or training programs in the IT and cybersecurity fields.

Creating a Flexible Learning Schedule: A personalized learning roadmap must include a flexible schedule. This schedule should accommodate other commitments while allowing consistent progress in learning. It should be realistic, considering the student's pace and lifestyle, and should be adaptable to changes in circumstances or learning needs. Keep in mind, it's not necessary to study/learn for hours a day. 30-minute learning sessions a day adds up to 14+ hours a month. Keep it manageable, but consistent.

Balancing Breadth and Depth of Knowledge: It's important to balance acquiring a broad understanding of IT fundamentals with specializing in specific areas. A well-rounded knowledge base helps in understanding how different components interconnect, while specialization allows for expertise in areas of interest. This balance builds a well-rounded skill set and can enhance career opportunities.

Documenting and Reflecting: Documenting the learning journey and reflecting on progress is instrumental for growth. This can be through maintaining a learning diary, portfolio, or a digital log. Regular reflection helps in recognizing achievements, understanding challenges, and adjusting the learning roadmap accordingly. It also serves as a motivational tool, showcasing how far you have come in your journey.

Contributions to Open-Source Projects

Open-source projects are collaborative efforts in software development where the source code is made freely available. These projects are pivotal to the IT community, fostering innovation and shared learning. They offer a platform where anyone, regardless of their professional background, can contribute to real-world software. Understanding how these projects operate and their ethos of collaboration and transparency is key for anyone looking to get involved. Open-source projects range from small programs to large-scale systems, and they often form the backbone of many commercial and non-commercial applications.

Benefits of Contributing: It's a practical way to apply and enhance your coding skills, learn new technologies, and understand real-world software development practices. Beyond skill development, it's a chance to give back to the community and help in the evolution of software that benefits many. Contributing can also significantly boost your professional profile, as it demonstrates your capabilities and commitment to potential employers. Additionally, it's an opportunity to work on projects that might be outside the scope of your regular job or studies, effectively broadening your experience.

Finding and Selecting Projects: Finding the right opensource project requires some research. Platforms like GitHub, GitLab, and Bitbucket host a multitude of projects. When selecting a project, consider your areas of interest and your current skill level. Look for projects that align with your professional goals and offer opportunities for learning and growth. It's also important to consider the project's community and support system. A welcoming and active community can greatly enhance your contribution experience and learning.

Making Meaningful Contributions: You can contribute through bug fixes, feature development, or even non-coding roles like writing documentation, designing user interfaces, or providing user support. Start with small, manageable tasks and gradually take on more complex challenges as you become more comfortable. Your contributions should aim to add value to the project, keeping in mind the project's goals and standards. Engaging actively with the community and participating in discussions can also lead to more impactful contributions.

Learning and Skill Development: Contributing to open-source projects is a powerful tool for learning and skill development. It exposes you to real-world coding practices and project management, often different from academic or isolated personal projects. You'll encounter new technologies and methodologies, which can be invaluable for your development. Additionally, reviewing code from others and having your code reviewed are great learning experiences. This practical exposure can significantly sharpen your problem-solving and technical skills.

Networking and Community Engagement: Active participation in open-source communities can offer excellent networking opportunities. You will get to interact with other developers, maintainers, and users from around the world. This network can become a valuable resource for career opportunities, professional advice, and mentorship. Engaging with the community also means participating in discussions, contributing to decision-making processes, and attending relevant meetups or conferences. This engagement can lead to lasting professional relationships and collaborations.

Building a Portfolio: Your contributions to open-source projects can serve as a tangible showcase of your skills for potential employers. A well-documented history of contributions reflects your technical abilities, collaborative skills, and dedication to continued learning. It's an effective way to build a portfolio that stands out, particularly in job interviews or when seeking career advancements. In the IT industry, where practical skills are highly valued, a portfolio enriched with open-source contributions can be a noteworthy differentiator.

Online Learning Platforms and Resources

Online learning platforms have revolutionized IT education, offering diverse and flexible learning experiences. These platforms range from massive open online courses like Coursera and Udemy to more specialized platforms like Pluralsight and Codecademy. Each platform has its unique strengths, whether it's in the breadth of courses offered, the depth of a specific IT domain, or the interactive nature of the learning experience. They provide students with the opportunity to study at their own pace and according to their own schedule, breaking down geographical and logistical barriers to education.

Types of Resources: These platforms offer a plethora of resources tailored to different learning styles and needs. Video tutorials provide visual and auditory learning experiences, while interactive coding exercises offer hands-on practice. Many platforms also feature discussion forums and community boards where learners can seek help, share insights, and discuss concepts. Additionally, some platforms offer quizzes and projects for practical application and assessment. This variety ensures that students can find resources that suit their preferred method of learning and engagement.

Choosing the Right Platform: Selecting the right online learning platform depends on individual learning goals, preferred content format, and specific IT topics of interest. If seeking a broad overview of IT, platforms with a wide range of courses might be ideal; i.e. Udemy, Coursera. For specialized skills, like data science or cybersecurity, platforms focusing on these niches would be more appropriate; i.e. TryHackMe, Hack the Box. It's also important to consider the quality of content, instructor expertise, and the learning community's engagement level. Free trials and sample courses can help in making an informed decision.

Cost-Effectiveness and Accessibility: One of the major advantages of online learning platforms is their cost-effectiveness. Many platforms offer free courses or are significantly less expensive than traditional education methods. This affordability makes IT education more accessible to a wider audience. Also, these platforms are accessible from anywhere with an internet connection, removing geographical constraints and offering flexibility that caters to different lifestyles and learning paces.

Staying Current with Industry Trends: The IT industry is fast-paced and constantly evolving. Online learning platforms are invaluable for keeping up-to-date with the latest trends, technologies, and best practices. They frequently update their course offerings to reflect current industry demands, providing learners with relevant and contemporary knowledge.

Community and Networking Opportunities: Many online learning platforms offer robust community features, providing opportunities for networking, collaboration, and peer-to-peer learning. These communities can be a great resource for problem solving, project collaboration, and building professional connections. They also offer a platform for students to share experiences, offer support, and gain insights from others who are on a similar learning journey. This communal aspect of online learning enriches the educational experience and can lead to valuable professional relationships.

As Chapter 3 concludes, we've navigated the diverse scenes of Information Technology certifications, practical learning, mentorship, and online resources. These elements are key stepping stones in your journey, offering a mix of theoretical knowledge and practical skills. Remember, the path in IT is not linear but a continuous cycle of learning, applying, and evolving. Your dedication to this journey will shape your success in this dynamic industry. Looking ahead, Chapter 4 will focus on translating your skills and experiences into a compelling professional narrative.

Reflect and Apply Questions

1. **Choosing Your Path:** Reflect on your interests in IT. How would you decide between a degree, certification, or bootcamp based on your career goals?
2. **Specialization vs. Generalization:** Considering the various IT specializations, which area interests you most, and why? What benefits come from a generalization of knowledge?
3. **Value of Certifications:** Think about certifications in IT. How could obtaining a specific certification impact your career prospects?
4. **Hands-On Experience:** Reflect on the importance of hands-on experience. How could engaging in practical projects or labs enhance your learning?
5. **Learning Roadmap:** Imagine creating your personalized learning roadmap. What key steps would you include to achieve your career aspirations?

Chapter 4: Crafting an Impressive Resume and Portfolio

In this chapter we discuss the aspects of crafting an impressive resume and portfolio, indispensable tools in your professional toolkit. We will guide you through creating a compelling narrative of your career, highlighting the importance of personal branding, and effectively showcasing your skills and experiences. Whether you're a newcomer to the field or an experienced professional seeking to advance, the insights provided here are designed to elevate your presentation to potential employers. We'll explore how to tailor your resume for roles, present your certifications and projects, and build a portfolio that speaks volumes about your capabilities. The goal is to equip you with the knowledge to create a professional image that stands out in the competitive IT job market.

Defining Your Brand and Presentation

This section encompasses the initial steps in crafting an impressive resume and portfolio. It covers defining your personal brand in IT and creating an engaging portfolio to showcase your skills and expertise. Additionally, it discusses the importance of consistent online presence and professional profiles, providing readers with strategies to establish a strong digital footprint.

Defining Your Personal Brand in IT

Personal branding in Information Technology is about creating a distinctive professional identity that encapsulates your skills, experiences, and values. It's how you present yourself to the industry and colleagues, both online and offline. This concept is valuable in a field as diverse and competitive as IT, where standing out can significantly impact career opportunities. A well-crafted personal brand helps in creating a memorable and impressive image that resonates with employers, clients, and peers. It's about communicating who you are, what you stand for, and what you can bring to the table in an IT role.

Identifying Your Unique Value Proposition: To establish a strong personal brand, start by identifying your unique value proposition (UVP). Your UVP is a clear statement that describes the unique benefit you offer, how you solve needs, and what distinguishes you from the competition. Consider your technical skills, work experience, problem-solving abilities, and any niche areas of expertise. Reflecting on feedback from colleagues and supervisors can also provide insights into your strengths and areas for improvement. Understanding and articulating your UVP is the first step of building a personal brand that genuinely represents your professional identity in IT.

Aligning Brand with Career Goals: Whether you aim to be a network security expert or a software development leader, your brand should reflect these aspirations. This alignment involves not only showcasing relevant skills and experiences but also engaging with the right communities, and creating content that resonates with your career path. It's about strategically positioning yourself in the areas that matter most to your career objectives, and ensuring that your personal brand communicates your career direction effectively.

Consistency Across Platforms: Personal branding requires consistency. Your professional image should be uniform across different platforms – from your LinkedIn profile to your personal website, and even your resume. This consistent representation builds recognition and trust. It includes the tone of your communication, the topics you talk about, and even visual elements like your profile picture or design choices. A consistent brand helps people remember you and reinforces your professional image every time they come across your profile or work.

Personal Branding and Professional Ethics: Your brand should reflect not only your skills and achievements but also your adherence to ethical standards. This might involve showcasing your commitment to data privacy, security practices, or inclusive technology design. Upholding ethical standards in your branding efforts enhances your credibility and trustworthiness in the industry.

Evolving Your Brand: Your personal brand should evolve as you progress in your career. As you acquire new skills, take on new roles, or shift your focus, your brand should reflect these changes. This evolution is a natural part of career growth and helps keep your professional image relevant and dynamic. Regularly revisiting and updating your brand ensures that it accurately represents your current status and aspirations.

Practical Examples of Personal Branding: To understand personal branding in action, look at IT specialists who have successfully established their brands. These individuals often have a consistent theme across their professional profiles, regularly contribute to industry discussions, and are known for specific expertise or viewpoints. They might be active on professional networks, run a tech blog, or speak at industry events, all of which reinforce their brand. Look on popular sites like LinkedIn or YouTube to find IT influencers/speakers. See how their brands compare from sites like above to their personal sites.

Crafting a Strong Resume Aligned with Your Brand

(*Find Resume Examples in Appendix D*)

Analyzing Job Descriptions: To craft a resume that stands out, start by thoroughly analyzing job descriptions. This involves identifying core skills, technologies, and qualifications that employers are looking for. Tailoring your resume to these specifics demonstrates that you are not only a suitable candidate but also attentive to details. Use this analysis to highlight aspects of your experience and skills that are most relevant to the job. This targeted approach can meaningfully increase your resume's impact, making it more appealing to potential employers.

Structuring Your Resume: A well-structured resume stands out in the IT industry. Begin with a clear header containing your contact information, followed by a professional summary that encapsulates your expertise and personal brand. Then, organize your resume into sections: Professional Experience, Education, Certifications, Skills, and Projects. Ensure the flow of information is logical and each section is easily navigable. Keep it clarified and concise; your resume should quickly convey your qualifications and fit for the role.

Highlighting Technical Skills and Proficiencies: In your resume, clearly highlight your relevant technical skills and proficiencies. This should include programming languages, software, tools, and systems you are adept in. Align these skills with your personal brand by focusing on those that best represent your strengths and career goals. Use specific examples or brief descriptions to demonstrate your proficiency level and experience with each skill. This section serves as a quick reference for employers to gauge your technical capabilities.

Incorporating Achievements and Certifications: When showcasing achievements and certifications, integrate them in a way that complements your personal brand. Mention certifications that are most relevant to the job you're applying for, and highlight achievements that demonstrate your expertise and contributions in previous roles. Use quantifiable metrics where possible, such as efficiency improvements or project outcomes, to give concrete evidence of your capabilities. This approach shows your qualify-cations and your impact in the field.

Balancing Technical and Soft Skills: A compelling IT resume strikes a balance between technical expertise and soft skills. Highlight your communication, problem-solving, and teamwork abilities alongside your technical skills. This demonstrates to employers that you can not only handle the technical aspects of the job but also collaborate effectively and adapt to various workplace situations. Include specific instances or projects where your soft skills played a substantial role in achieving results.

Customizing for Different Roles: Customizing your resume for different IT roles will help you stand out. For a developer position, emphasize your coding skills and development projects. For a network administrator role, focus on your experience with network infrastructure and security. Tailor the emphasis of your resume according to the role's requirements while supporting the core elements of your personal brand. This customization shows that you understand the nuances of the role and are a suitable fit for the specific position. Popular jobs will have hundreds of applicants sending their basic application.

Creating an Engaging IT Portfolio

An IT portfolio is a comprehensive collection of work that showcases your technical skills, projects, and accomplishments. It goes beyond the resume to provide a visual and detailed insight into your professional journey. A well-crafted portfolio highlights your range of abilities, from technical expertise to problem-solving and project management skills. It serves as a dynamic tool for potential employers or clients to gauge the depth and breadth of your experience. A portfolio can be a powerful way to demonstrate how your skills have evolved and been applied in various contexts.

Benefits of a Portfolio: It acts as a tangible showcase of your technical prowess and project involvement, illustrating your ability to apply skills in real-world scenarios. A portfolio can differentiate you from other candidates by providing a deeper, more engaging look at your work and achievements. It's particularly beneficial in the IT field, where practical skills and project outcomes are highly valued. Additionally, a portfolio can be a great conversation starter and reference point during interviews.

Selecting Portfolio Projects: It is advised to choose the right projects for your portfolio. Focus on including work that showcases a variety of skills and reflects your career highlights. Select projects that demonstrate your problem-solving abilities, technical expertise, and creativity. It's also important to include projects that align with your career aspirations and personal brand, as these will resonate most with your target audience. The goal is to create a diverse and impactful portfolio that captures the essence of your professional journey.

Describing Projects Effectively: Each project description should clearly articulate the problem, your role, the solutions you implemented, and the results achieved. Use language that conveys your technical expertise while being accessible to non-specialists. This approach showcases your ability to communicate complex ideas effectively. Including measurable outcomes or success metrics can further enhance the impact of your descriptions.

Portfolio Design and Layout: The design and layout of your portfolio are as important as the content. It should be user-friendly, visually appealing, and reflect your professional brand. A clean, organized layout makes it easier for viewers to navigate and understand your work. If you're not a designer, consider using portfolio templates or platforms that provide a professional look and feel. The design should enhance your work, not overshadow it.

Online Portfolios and Personal Websites: These are invaluable in today's digital world. It offers a convenient, accessible way for potential employers and clients to view your work. An online portfolio can include interactive elements, links to live projects, and downloadable resources. It also allows you to reach a wider audience and can be easily shared. Ensure your online presence is professional, up-to-date, and consistent with your personal brand

Consistent Online Presence and Professional Profiles

A unified professional image across online platforms is important to establish a credible IT persona. Your online profiles should consistently reflect your professional skills, achievements, and personal brand. This uniformity in presentation enhances recognition and reinforces your reputation in the industry. Ensure that your LinkedIn, GitHub, personal website, and other professional profiles all convey the same image and message. A cohesive online presence builds your professional identity and makes you more recognizable and trustworthy to potential employers.

Best Practices for Professional Networks: LinkedIn and other professional networks are excellent for showcasing your IT career and network. Optimize your LinkedIn profile by highlighting your skills, experiences, and achievements. Be sure it aligns with your resume and portfolio. Regular updates, endorsements, and recommendations enhance your profile's credibility. Engage with the community by sharing insights, commenting on industry trends, and connecting with fellow specialists. These platforms are not just about displaying your career history; they're also tools for engaging with and contributing to your professional community.

Managing Digital Footprints: Your digital footprint plays a significant role in how you're perceived professionally. Regularly review and curate your online presence, ensuring that all public information is professional and reflects your personal brand. Be mindful of your contributions on social media, blogs, and forums. It is recommended to remove or update any outdated or improper content. Remember, potential employers and colleagues are likely to research your online presence, so be sure that it accurately represents your professional demeanor.

Engaging with Online Communities: Active participation in online IT communities can significantly boost your profile. Contribute to discussions, offer solutions, and share your knowledge on platforms like Stack Overflow, Reddit's IT communities, or specialized tech forums. This engagement demonstrates your expertise and commitment to the field. It also offers opportunities for learning, networking, and staying updated with industry trends. Engaging in these communities can elevate your professional standing and can often lead to valuable connections and opportunities.

Real-World Example: Jake, an aspiring data analyst, regularly engaged in online IT communities. He often contributed to discussions on platforms like GitHub and Stack Overflow. Jake helped solve complex data problems and shared his own projects for community feedback. When he applied for a data analyst position, he included these contributions in his resume. During his interview, he was able to discuss these experiences, demonstrating his proactive engagement and continuous learning in the field. This example illustrates the value of active participation in IT communities and how it can enhance a professional profile.

Leveraging Social-Media for Professional Growth: Social-media can be a powerful tool for professional growth when used strategically. Platforms like X (Twitter), LinkedIn, and even Instagram can be leveraged to showcase your IT expertise, share industry news, and connect with thought leaders. Use these platforms to demonstrate your knowledge, engage in conversations, and share your professional achievements. A proactive approach to social media can enhance your visibility in the field and open doors to new opportunities and networks. Remember to maintain professionalism and align your social media activities with your overall personal brand strategy.

Showcasing Your Credentials and Adaptability

Transitioning into the practical aspects of resume building, this section focuses on showcasing certifications, projects, and experiences effectively. It includes tips for incorporating certs and projects into your resume and portfolio. Also, it addresses the importance of adapting your resume and portfolio over time to reflect your evolving skills and career goals.

Displaying IT Certifications and Projects in Your Resume

Choose certifications that align closely with the job requirements and your career goals. Highlight those that are most recognized and respected in the IT industry, as they carry more weight. Place them prominently on your resume to catch the attention of recruiters and hiring managers. The relevance and prestige of certifications can significantly enhance your candidacy.

When detailing projects, focus on those where you had a significant role and which align with your career trajectory. Describe the technologies used and your specific contributions. Highlight the project's outcomes and how they benefited the team or company. This approach showcases your practical experience and problem-solving skills.

Quantify your achievements in projects wherever possible. Use specific metrics like increased efficiency, reduced costs, or improved performance. This quantification supplies concrete evidence of your skills and impact. It makes your contributions tangible and more impressive to potential employers.

Seamlessly integrate your certifications and projects into your resume in a way that reinforces your personal brand. Use them to demonstrate your expertise and areas of specialization. Ensure they reflect the narrative of your professional development and aspirations. This integration strengthens your personal brand and makes your resume more compelling.

Regularly update your resume with new certifications and recent projects. This ensures your resume reflects your current skill set and latest achievements. Keep the content relevant to the evolving IT industry and the roles you are targeting. A current and relevant resume demonstrates your commitment to professional growth.

Effective Cover Letters as a Branding Tool

In the IT job application process, a cover letter is more than a formality; it's an opportunity to narrate your professional story. It complements your resume by providing context and personality to your application. A well-crafted cover letter should not only summarize your skills and experiences but also articulate how and why you fit the specific role and company. It's your chance to make a memorable first impression, showing your enthusiasm and fit for the role beyond what a resume can convey.

Personalizing Your Cover Letter: It is recommended to personalize your cover letter for each job application. This will demonstrate your genuine interest in the specific role and company. Tailor your letter to address the unique requirements and culture of the company, and explicitly mention how your skills and experiences align with the job description. This targeted approach shows the employer that you've done your research and are serious about the position.

Real-World Example: Alex is an IT expert, who applied for a network engineer position. Alex meticulously researched the company and discovered they were expanding their cloud infrastructure. In his cover letter, Alex highlighted his recent experience with cloud migration projects and his certifications in cloud technologies. He also mentioned a specific challenge he overcame during a migration project, showcasing his problem-solving skills and direct relevance to the company's current needs. This personalized approach caught the hiring manager's attention, leading to an interview where Alex could further elaborate on his experiences. This example illustrates the impact of a well-tailored cover letter that resonates with an employer's specific needs and projects.

Conveying Your Personal Brand: Your cover letter is the prime opportunity to communicate your personal brand. Use it to express your professional identity, values, and what sets you apart from other candidates. It should reflect your unique blend of skills and experiences and how they equip you to excel in the prospective role. This personalized narrative can make your application stand out in a sea of generic letters.

Highlighting Relevant Projects and Skills: Use your cover letter to spotlight projects and skills that are particularly relevant to the job. Discuss specific experiences that demonstrate your capabilities and directly relate to the role's requirements. This approach not only illustrates your qualifications but also shows your ability to apply your skills in practical scenarios. Be concise yet impactful in showcasing these highlights.

Real-World Example: Consider Sarah, a software developer who applied for a senior position. In her resume, Sarah emph-asized a recent project where she led a team to develop a mobile application using Agile methodologies. She detailed her role in the project, the challenges they faced, and the innovative solutions they implemented, which resulted in a 30% increase in app efficiency. By quantifying the impact of her leadership and technical skills, Sarah effectively showcased her suitability for a senior role.

Structure and Tone: Start with a strong introduction, followed by a body that elaborates on your qualifications, and a concise conclusion. The structure and tone of your cover letter should be professional yet approachable. Maintain a tone that is confident and enthusiastic without being overbearing. The letter should be easy to read, with clear paragraphs and a logical flow.

Call to Action and Closing: End your cover letter with a compelling call to action and a professional closing. Encourage the employer to contact you for further discussion or an interview. Express your appreciation for their consideration and reiterate your enthusiasm for the opportunity. A strong closing leaves a lasting impression and can prompt the employer to move forward with your application.

Adapting Your Resume and Portfolio Over Time

It's important to update your resume and portfolio regularly to reflect your evolving skill set and professional experiences. As you acquire new certifications, complete significant projects, or gain new skills, incorporate these into your documents. This practice ensures that your professional presentation remains up-to-date and showcases your continuous learning and growth.

Incorporating New Technologies and Trends: Stay informed about the latest technologies and industry trends, and reflect these in your resume and portfolio. Showcasing your knowledge and experience with emerging technologies can make you more attractive to potential employers. This could involve updating your skills section, adding new projects, or even revising the language to include current industry buzzwords. At the time of writing, popular technologies include AI and Blockchain. If you can get experience via hands on learning or coursework, these will help your resume stand out. Keeping your documents current with industry developments illustrates your adaptability and eagerness to stay ahead in a rapidly evolving field.

Re-evaluating Career Goals and Objectives: Periodically reassess your career goals and objectives to ensure your resume and portfolio align with them. As your career path evolves, so should the content of your professional documents. This might mean emphasizing different aspects of your experience, focusing on new skills, or even shifting the overall narrative of your career story. An updated resume and portfolio that reflects your current career aspirations will more effectively help you reach your next professional milestone.

Feedback and Continuous Improvement: Continuously seek feedback from mentors, peers, and industry experts. Use this feedback to make improvements and refinements. This process of regular review and adaptation will help maintain a high-quality, impactful professional presentation. Feedback can provide new perspectives and ideas, helping you to create documents that resonate better with your target audience.

As we wrap up Chapter 4, we've navigated the nuances of crafting a resume and portfolio that not only showcase your technical skills and projects but also reflect your unique personal brand. We've seen how a well-structured resume, a compelling portfolio, and a consistent online presence can significantly elevate your professional image in the IT world. The integration of real-world examples emphasized the practical application of these concepts. Your resume, portfolio, and online profiles are dynamic tools; they evolve as you grow in your career and should be regularly updated to reflect your latest achievements and learning. Looking ahead, Chapter 5 will shift our focus to the strategies and techniques for effectively searching for and securing job opportunities in the field.

Reflect and Apply Questions

1. **Branding Yourself:** How would you define your personal brand, and how can you reflect it in your resume and portfolio?
2. **Resume Tailoring:** Think about a job you'd like to apply for. How would you tailor your resume to align with that job's requirements?
3. **Portfolio Selection:** Consider the projects you've worked on. Which would you choose for your portfolio to best showcase your skills?
4. **Online Presence:** Reflect on your current online presence. What changes would you make to enhance your professional image?
5. **Continuous Improvement:** How would you gather and implement feedback to continuously improve your resume and portfolio?

Chapter 5: Navigating the Job Market

In this chapter, we dive into the vigorous world of the IT job market. As the landscape of technology continues to evolve, so do the opportunities and challenges within the job sector. We'll explore aspects like the impact of remote work, the rise of emerging technologies, and how they influence job roles and demand. We'll also examine strategies for adapting to rapid technological changes and shifts in IT specializations, providing insights into globalization and its effects on the workforce. This chapter aims to equip you with the knowledge and tools necessary to navigate this complex market, adapt to its changes, and find your place in the global community.

Understanding the Landscape

In this section, we explore the intricacies of the job market, providing insights into industry trends. By understanding the landscape of the industry, you'll be better equipped to navigate job opportunities and make informed decisions about your career path.

Industry Insights

Impact of Remote Work: Companies are now more open to hiring IT specialists from a wider geographic area, increasing opportunities but also competition. This shift also emphasizes the need for skills in remote collaboration and cloud-based tools. IT roles are adapting to support remote infrastructure, cybersecurity for distributed networks, and online project management. The remote work trend offers flexibility but also requires individuals to be self-motivated and adept at managing work-life boundaries.

Rise of Technologies and Their Influence: Emerging technologies like AI, cloud computing, and IoT are creating new job roles and altering existing ones in the industry. There's an increasing demand for professionals skilled in these areas, leading to the creation of roles like AI specialists and cloud architects. Staying updated with these technologies will support career progression and relevance in the job market. IT pros must continuously learn and adapt to stay ahead in this consistently evolving landscape.

Demand Shifts in IT Specializations: Cybersecurity, data science, and software development are currently in high demand.

Specialists in these areas are seeing more job opportunities and potentially higher salaries. Understanding these trends can help individuals make informed decisions about their career paths and professional development focus.

Globalization of the Workforce: Companies are increasingly open to hiring international talent, which offers a broader range of opportunities but also introduces global competition. This trend has implications for cultural adaptability, communication skills, and the ability to work across time zones. IT prospects must be prepared to operate in a global work environment.

Real-World Example: Consider the case of Anna, an IT expert specializing in traditional database management. With the growing dominance of cloud technologies, Anna recognized the need to adapt her skill set. She undertook training in cloud-based data management and analytics, gaining proficiency in platforms like AWS and Azure. This proactive approach safeguarded her career against obsolescence and opened new opportunities in cloud data services. Anna's story highlights the importance of adapting to technological changes in the industry and how continuous learning can lead to career growth and stability.

Understanding the Job Market

The IT job market is dynamic, with certain skills and roles consistently in high demand. Areas like cloud computing, data science, and cybersecurity are seeing substantial growth. Understanding these trends is crucial for IT experts looking to advance or pivot their careers. Keeping abreast of such trends helps in aligning your skill set with market demands, enhancing employability. Additionally, soft skills like project management and effective communication are increasingly valued. Recognizing and adapting to these trends can substantially impact your career trajectory in the industry.

Geographic and Industry Variations: Job opportunities and demands can vary significantly across different regions and industries. For instance, tech hubs like Silicon Valley offer a concentration of opportunities in innovative tech companies, whereas other regions may have a focus on IT roles in government or healthcare sectors. Understanding these geographical and industrial variations is important when considering relocation or targeting specific sectors. Tailoring your skill set and job search strategies to align with these variations can greatly enhance your chances of success.

Impact of Emerging Technologies: Advancements in technologies such as AI, cloud computing, and cybersecurity are creating new job roles and transforming existing ones. For example, the rise of cloud technology has led to a surge in demand for cloud architects and engineers. Similarly, the growing importance of data protection has elevated the need for cybersecurity experts. Staying informed about these technological shifts is beneficial to finding new career opportunities and ensuring your skills remain relevant.

Real-World Scenario: Consider the case of Emily, an IT professional with a background in traditional networking. With the rise of cloud computing, she noticed a shift in the industry's focus. To adapt, Emily pursued certifications in cloud technology and gained hands-on experience through personal projects. Her proactive approach allowed her to transition into a cloud solutions architect role and greatly increased her marketability and career prospects.

Specialization vs. Generalization: In the IT job market, there's an ongoing debate about the merits of specialization versus generalization. Specialization entails becoming deeply proficient in a particular area of IT, such as cybersecurity, cloud computing, or data analytics. Specialized professionals often command higher salaries and enjoy greater demand in industries where their expertise is valued. For example, a cybersecurity specialist with certifications like Certified Information Systems Security Professional (CISSP) or Certified Ethical Hacker (CEH) might find lucrative opportunities in industries prioritizing robust security measures. However, specialization can sometimes lead to pigeonholing, limiting career flexibility and adaptability to evolving industry trends.

On the other hand, generalization involves acquiring a broad skill set across multiple domains, enabling specialists to adapt to diverse roles and responsibilities. Generalists possess a versatile toolkit that allows them to pivot between different projects and industries, making them valuable assets in lively environments where versatility is prized. For instance, a generalist IT consultant might excel in roles requiring a blend of technical expertise, project management skills, and business acumen. Yet, generalization may entail sacrificing in-depth knowledge in specialized areas, potentially limiting advancement opportunities in fields where specialization is critical.

Ultimately, the decision between specialization and generalization hinges on individual preferences, career goals, and market dynamics. Some experts opt for a hybrid approach, combining specialized expertise in one or two areas with a broad foundation across multiple domains. This hybrid model allows individuals to leverage their specialized skills while remaining adaptable to diverse opportunities. In today's rapidly evolving IT landscape, staying well-informed of industry trends and continuously honing skills—whether through specialization, generalization, or a combination of both—is essential for long-term success.

Adapting to the Changing IT Landscape: Professionals must be proactive in upgrading their skills and staying informed of new technologies and methodologies. This might involve formal education, online courses, or self-directed learning. Staying flexible and open to change is key to maintaining relevance in the IT field. Those who can adapt to new technologies and industry shifts are more likely to thrive and advance in their careers.

Strategies for Success

This section focuses on essential strategies for success in the job market. From effective job search techniques to crafting targeted applications, I will provide actionable advice to help you stand out and secure your desired IT role. By implementing these strategies, you'll enhance your chances of success in a competitive job market.

Job Search Strategies for IT Professionals

Online job boards and websites are increasingly common in the modern job search, especially in IT. Platforms like LinkedIn, Indeed, and specialized IT job boards offer a wide range of opportunities. They allow you to filter job searches by location, skill set, and experience level, making it easier to find relevant positions. To maximize their potential, keep your profiles updated and tailor your applications to each job. Regularly engaging with these platforms can increase your visibility to recruiters and alert you to the latest job openings in your field.

Leveraging Recruitment Agencies and Headhunters: Both can be invaluable resources in your job search. Recruiters and headhunters often have access to job openings that aren't publicly advertised and can match you with positions that align with your skills and career goals. Building a good relationship with a recruiter can provide you with industry insights and advice on how to improve your employability. It's important to choose agencies or headhunters with a strong track record in the IT sector for the best results.

The Power of Networking in Job Searches: Networking plays a pivotal role in uncovering job opportunities. Building a professional network through industry events, conferences, and online platforms can open doors to job opportunities that are not advertised. Networking is not just about finding a job; it's about building relationships that can provide support, advice, and insights into the industry. Engaging with your network regularly and contributing value can significantly enhance your job search prospects.

Real-World Scenario: John, an IT specialist, was looking for a new role. He regularly attended local tech meetups and engaged in online forums. At one such event, he had an informal chat with a company executive about industry trends, which led to a discussion about his career aspirations. Impressed by John's knowledge and enthusiasm, the executive referred him for an open position at her company, where John eventually got the job. This example highlights how effective networking can lead to unexpected and valuable job opportunities.

Utilizing Company Websites and Career Pages: Directly approaching companies through their career page is a proactive job search strategy. Many companies list job openings on their websites before posting them on job boards. This method allows you to learn in-depth information about the company and tailor your application to their specific needs and culture. Additionally, applying through a company's career page often ensures your application is directly reviewed by their HR department.

Creating Job Alerts and Staying Informed: Setting up job alerts on various job boards and company websites is an efficient way to stay informed about new openings. Customize these alerts based on your preferred job titles, companies, and locations. This approach ensures you receive prompt notifications about relevant job opportunities, allowing you to apply immediately. Organizing your job search and keeping track of applications can also enhance your efficiency and effectiveness in the job search process.

Alternative Job Search Methods: Beyond traditional job boards and networking, consider alternative methods like attending job fairs and industry conferences. These can provide opportunities to meet potential employers face-to-face. Utilizing social media platforms, especially professional ones like LinkedIn, can also uncover job opportunities. Participating in community projects, online forums, or contributing to open-source projects can also draw attention to your skills and lead to job offers.

Leveraging Networking in Job Searches

Networking is a key element in the IT job search process, often leading to opportunities that aren't publicly advertised. Many jobs are filled through referrals and personal connections, making your professional network an invaluable resource. Building relationships with industry professionals can open doors to these hidden job markets. Actively engaging in discussions, attending industry events, and taking part in professional groups can significantly increase your visibility and chances of being referred for a job. Remember, networking is about building genuine relationships, not just collecting contacts.

Online Networking for Job Searches: In today's digital age, online networking plays a big role in job searches. Platforms like LinkedIn are valuable for connecting with industry leaders, joining relevant IT groups, and engaging in professional discussions. Your online presence can showcase your skills, interests, and professional aspirations to potential employers. Actively participating in discussions, sharing relevant content, and connecting with people in your desired field can lead to job leads and referrals. Make sure your online profiles are up-to-date and reflect your professional brand.

Maximizing Networking Events for Job Hunting: Networking events, conferences, and workshops offer a direct route to connect with industry experts and potential employers. When attending these events, have a clear strategy: know what types of roles you're interested in and which companies or individuals you want to connect with. Prepare a brief but effective introduction that summarizes your background and career interests. Engage in meaningful conversations, showing genuine interest in others' work and sharing insights from your own experience. Collect contact information and make notes about your conversations for effective follow-up.

Follow-Up Strategies Post-Networking Events: Send personalized LinkedIn requests or emails, referencing specific details from your conversations. Express gratitude for their time and reiterate your interest in any opportunities they mentioned. Regularly check in with your contacts, share relevant articles or information, and keep them updated on your professional progress.

This consistent and thoughtful follow-up keeps the relationship alive and can lead to job opportunities down the line.

Crafting Targeted Job Applications

When applying for IT jobs, the first step is to thoroughly analyze the job description. Pay close attention to the listed requirements and preferred qualifications. This analysis helps you understand exactly what the employer is looking for and which of your skills and experiences are most relevant. Use this information to tailor your application, ensuring it addresses the specific needs and challenges the employer is looking to solve. This targeted approach demonstrates your attention to detail and increases the relevance of your application.

Tailoring Your Resume and Cover Letter: Customize your resume and cover letter for each job application. Reflect on the job description and emphasize the aspects of your experience that align with the role's requirements. This could mean highlighting certain projects, skills, or experiences more prominently, depending on what the job entails. Tailoring your application shows that you have taken the time to understand the role and believe in your fit for the position, making your application stand out.

Incorporating Keywords for Applicant Tracking Systems (ATS): Many companies use Applicant Tracking Systems to screen resumes. These systems scan for specific keywords related to the job. These keywords often scan for required certs, degrees, or skills. To ensure your application makes it through these filters, include relevant keywords from the job description in your resume and cover letter. However, it's important to do this naturally and truthfully, ensuring your skills and experiences genuinely align with the keywords you include. If the job description specifies preferred certifications that you may have, like the Security+ for instance, be sure to include that on your resume.

Demonstrating Company-Specific Knowledge and Interest: Showcasing your knowledge about the company and expressing genuine interest can set your application apart. Research the company's values, recent projects, and challenges in the industry. Reflect this knowledge in your application, explaining why you're not just interested in the role but also enthusiastic about contributing to the company specifically. Spending 30 minutes researching a company's values and direction could be the key to securing the job offer. Employers value candidates who are knowledgeable and excited about their organization.

Real-World Example: An IT professional, Linda, applied for a software developer position at a tech startup focused on healthcare technology. In her cover letter, she referenced the company's recent move into telehealth solutions, aligning it with her previous experience in developing healthcare apps. She also expressed her passion for leveraging technology to improve patient care, mirroring the company's mission.

This demonstration of company-specific knowledge and interest made Linda's application stand out. Her cover letter resonated with the hiring team as it showed her alignment with the role and with the company's vision and goals. It resulted in her being shortlisted for an interview, where she could further elaborate on her experiences and ideas relevant to the company's projects.

Showcasing Transferable Skills: When applying for various IT roles, especially those slightly different from your previous positions, it's important to highlight transferable skills. Focus on the skills that are relevant across different roles, such as problem-solving, project management, or technical adaptability. Illustrate these skills with examples from your past experiences, showing how they can be applied effectively in the new role. This approach can make you a compelling candidate even for positions that are not an exact match with your previous roles.

Follow-Up: Following up on job applications can demonstrate your interest and initiative. However, it's important to do so respectfully and professionally. Wait for a reasonable amount of time after submitting your application, typically one to two weeks, before sending a polite follow-up email. In your email, reiterate your interest in the position and inquire about the status of your application. This can help keep your application top-of-mind and show your enthusiasm for the opportunity.

Preparing for Advancement

Next, we will discuss interview prep and negotiating job offers. We will then dive into evaluating job fit and company culture, followed by career progression and continuous learning.

Preparing for IT Job Interviews

IT interviews can vary widely, ranging from technical assessments to behavioral interviews and case studies. Technical interviews often focus on specific knowledge or problem-solving skills, while behavioral interviews assess how you handle certain situations. Case study interviews might require you to solve a hypothetical industry-specific problem. Understanding the format of the interview you're facing is crucial for effective preparation. It's beneficial to clarify with the recruiter beforehand the type of interview to expect.

Technical Interview Preparation: Preparing for technical interviews requires a solid understanding of concepts and technologies relevant to the job. Review your technical knowledge, and if possible, practice coding or technical problem-solving exercises. Familiarize yourself with common types of questions or challenges specific to your field. Hands-on practice, whether through personal projects, simulations, or lab work, is invaluable for demonstrating real-world application of your skills.

Behavioral Interview Techniques: In behavioral interviews, employers are looking to assess your soft skills and how you handle workplace situations. Utilize the STAR method (Situation, Task, Action, Result) to structure your answers, providing concrete examples that illustrate your skills and experiences. Ideally, these examples should reflect the competencies the job requires. Practic-ing your responses to common behavioral questions can make you more comfortable and confident during the actual interview.

Researching the Company and Interviewers: Before the interview, research the company thoroughly. Understand its products, services, culture, and recent news. If possible, learn about your interviewers' backgrounds. Check their LinkedIn profile and see what their role is and how long they have been with the organization. This knowledge allows you to tailor your responses to align with the company's values and needs, and it demonstrates your genuine interest in the position and the organization.

Mock Interviews and Practice: Mock interviews are an excellent way to practice and refine your interviewing skills. They can help identify areas for improvement and reduce anxiety. You can conduct mock interviews with a mentor, peer, or through professional services. Receiving feedback on your answers, body language, and overall presentation can be beneficial in preparing for the actual interview.

Questions For Interviewers: Prepare thoughtful questions to ask your interviewers. These questions should demonstrate your understanding of the role and your interest in the company. Inquire about specific projects, team dynamics, growth opportunities, or the company's future plans. Thoughtful questions can show your enthusiasm for the role and provide valuable insights for your decision making.

Handling Virtual Interviews: Virtual interviews have become a norm, and it's important to master them. Ensure your technology works flawlessly, and your environment is professional and distraction free. Test your camera and microphone beforehand, and familiarize yourself with the software being used. Dress professionally, maintain eye contact, and be mindful of body language even in a virtual setting.

Post-Interview Etiquette: After the interview, send a thank-you email to your interviewers. Express appreciation for their time and reiterate your interest in the role. This follow-up is not only polite but can also keep you top of mind. If you haven't heard back within the timeframe provided, it's appropriate to send a polite follow-up email to inquire about the status of your application.

Importance of Hands-On Practice: Real-world experience and hands-on practice can be a game changer in these interviews. Discussing challenges you've faced and resolved in practical settings, such as a home lab, can significantly showcase your technical acumen and problem-solving skills. It demonstrates not just theoretical knowledge, but also how you apply it in real scenarios. Such examples can vividly illustrate your competence and resourcefulness, traits highly valued in IT professionals. Whether it's troubleshooting network issues, coding a complex program, or setting up a virtual server environment, these experiences provide concrete evidence of your skills and your ability to handle real-world IT challenges.

A Personal Example: During an interview, I recall discussing a problem I encountered in my home lab. I made a configuration error on a network router, which led to a series of troubleshooting steps. Describing how I identified and resolved the issue not only showcased my technical skills but also my methodical approach to problem-solving. This story resonated with the interviewers as it highlighted my hands-on experience and ability to learn from mistakes, key traits in Information Technology.

Negotiating Job Offers

An IT job offer includes several components beyond the base salary. It's vital to understand each element, such as health benefits, retirement plans, bonuses, stock options, work hours, remote work flexibility, and opportunities for advancement. Each of these factors can play a role in your overall job satisfaction and should be carefully considered when evaluating an offer. Understanding the complete package will help you make a well-informed decision and provide a basis for any negotiations.

Researching Market Rates and Salary Benchmarks: Before entering negotiations, research the market rates and salary benchmarks for your role in the industry. Utilize resources like salary surveys, industry reports, and professional forums to gather information. Understanding what specialists with similar skills and experience are earning in your region will help you negotiate a competitive salary. This research provides you with data to back up your salary requests during negotiations.

Evaluating the Entire Offer Package: Look beyond the salary when evaluating a job offer. Consider the benefits package, which can include health insurance, retirement plans, paid time off, and other perks. Assess the work-life balance, company culture, and opportunities for professional growth. Sometimes, a slightly lower salary might be offset by a great benefits package or a positive work environment.

Effective Negotiation Strategies: When negotiating salary, be clear, confident, and assertive, but also reasonable and respectful. Start by expressing your enthusiasm for the position and the company. Then present your case, backed by your research on market rates and your professional value. Be prepared to explain why you believe the salary should be higher, citing specific examples of your skills and achievements.

Real-World Example: Emily, a project manager, received a job offer with a salary lower than her expectation. She conducted thorough market research and found that her desired salary was aligned with industry standards for her experience and region. During the negotiation, Emily presented this data and articulated her proven track record of successfully managing complex projects. Impressed by her preparation and understanding of her worth, the company revised their offer, meeting her salary expectations.

Negotiating Benefits and Other Aspects: Negotiations can also include non-monetary aspects. If the salary offer is firm, consider negotiating for additional vacation time, flexible working hours, a professional development allowance, or a sign-on bonus. These elements can meaningfully enhance your overall job satisfaction and work-life balance.

Handling Multiple Job Offers: If you have multiple job offers, approach the situation tactfully. You can use the offers as leverage in negotiations, but do so respectfully and without pressuring the employers. Be honest and transparent about your situation, and consider each offer carefully based on all aspects, not just the salary.

Communicating Professionally: Throughout the negotiation process, maintain a professional and respectful tone. Avoid making demands; instead, frame your requests as part of a discussion about finding a mutually beneficial agreement. Clear and open communication can lead to positive outcomes and maintain good relationships with potential employers.

Decision Making and Accepting the Offer: Once you've negotiated and received the final offer, take the time to make a well-informed decision. Consider all aspects of the offer and how they align with your career goals and personal needs. If you decide to accept, communicate your acceptance clearly and professionally, and ask for the offer in writing. This step ensures that both you and the employer have a clear understanding of the agreed-upon terms.

Evaluating Job Fit and Company Culture

Researching a company's culture is important for evaluating job fit. Review the company's website, social media, and any published materials to understand their values and work environment. Look for insights into their approach to teamwork, innovation, employee engagement, and diversity. This initial research can give you a sense of whether the companies environment aligns with your preferences and working style.

Assessing Alignment with Personal Values: It's important that your personal values align with the company's culture. Reflect on what matters most to you in a workplace – whether it's innovation, integrity, collaboration, or work-life balance. Compare these values to what you've learned about the company to gauge if there's a good fit. This alignment will help foster long-term job satisfaction and success.

Importance of Work-Life Balance: Evaluate the company's stance on work-life balance. Look for policies on flexible working hours, remote work options, and the amount of overtime typically expected. Understanding this aspect of the company culture will help you make an informed decision, especially if maintaining a certain work-life balance is important to you.

Career Development Opportunities: Investigate the organization's commitment to employee growth and development. This could be through training programs, mentorship opportunities, or clear pathways for career advancement. A company that invests in its employees' growth is likely to support your long-term career aspirations.

Feedback and Reviews: Check employee reviews on platforms like Glassdoor or seek out feedback from current or former employees. These insights can be invaluable in understanding the real work environment and culture of the company. However, take these reviews with a grain of salt and look for consistent patterns rather than isolated comments.

Asking the Right Questions: During interviews, ask questions about the company culture, team dynamics, and management style. Inquire about what success looks like in the role and how the company supports employee development. The answers to these questions can offer insights into whether the company is a good fit for you.

Career Progression and Continuous Learning

Setting Long-Term Career Goals: In a field characterized by rapid change and advancement, establishing long-term career goals is crucial. Goals provide direction and motivation, helping you to focus your efforts and resources effectively. They should be specific, measurable, achievable, relevant, and time-bound (SMART). Reflect on where you want to be in the next five, ten, or fifteen years, considering roles, technologies, and specializations that interest you. Regularly revisit and adjust these goals as you progress in your career and as the industry evolves.

Identifying Opportunities for Career Advancement: Career advancement often requires proactivity and awareness of opportunities within your current organization and the industry at large. Stay alert to internal job postings, project opportunities, and potential mentorships. Outside your organization, keep an eye on industry trends and job openings that align with your respective career goals. Networking, ongoing education, and a willingness to step out of your comfort zone are key to capitalizing on these opportunities.

The Role of Continuous Learning and Certifications: In the IT industry, where technologies and best practices evolve rapidly, continuous learning is not just beneficial—it's a requirement. It's what differentiates proactive IT professionals from those who merely keep pace. Engaging in ongoing education, whether through formal courses, certifications, workshops, or self-directed learning, ensures you remain relevant and competitive. For instance, certifications in emerging fields like cloud computing or cybersecurity can open new career avenues and demonstrate your commitment to staying at the forefront of technology. Moreover, continuous learning shows potential employers that you're adaptable and willing to invest in your professional development.

Real-World Example: Consider the story of Eric, a dedicated IT expert specializing in software development. Early in his career, Eric focused mainly on Java development. However, as the industry started shifting towards cloud technologies and DevOps practices, he realized the need to expand his skill set. Eric invested time in learning cloud platforms like AWS and Azure, and gained proficiency in DevOps tools. His initiative to learn and adapt made him an invaluable asset to his team and led to a promotion. He transitioned from a software developer to a lead DevOps engineer in his organization. Eric's story shows the tangible benefits of continuous learning in the field, where staying updated with the latest technologies can lead to substantial career advancement.

Adapting to Technological Changes: The IT industry is constantly evolving, with emerging technologies like blockchain, the Internet of Things (IoT), and cybersecurity reshaping the landscape and creating new career opportunities. Let's briefly dig deeper into each:

- **Blockchain Technology**: Initially recognized for its application in cryptocurrencies, blockchain technology is now finding its way into diverse sectors such as finance, supply chain management, and healthcare. IT experts looking to specialize in blockchain need to understand distributed ledger technology, smart contracts, and consensus algorithms. As companies increasingly adopt blockchain solutions, opportunities for roles like blockchain developers and solution architects are on the rise.
- **Internet of Things (IoT)**: IoT is transforming how we interact with devices and data, creating a more connected world. This has led to a surge in demand for professionals skilled in sensor technologies, data analytics, and network security tailored to smart devices. Roles in IoT project management, security analysis, and system design are

becoming more prevalent. By learning these skills, IT specialists can position themselves at the forefront of this expanding field.

- **Cybersecurity**: With the ever-increasing amount of data and its associated risks, cybersecurity has become a critical concern for all organizations. This has led to a high demand for IT professionals skilled in areas like threat intelligence, penetration testing, and security policy development. Pursuing certifications and training in cybersecurity can open doors to roles such as cybersecurity analyst, security architect, or compliance officer.

Incorporating these technologies into your skill set requires a commitment to continuous learning and a keen eye for industry trends. Staying ahead in these areas can lead to exciting and rewarding career paths that are aligned with the future direction of the industry.

Seeking Feedback and Conducting Self-Assessments: Regular feedback and self-assessment play a major role in professional growth. Seek constructive feedback from supervisors, mentors, and peers to gain insights into your strengths and areas for improvement. Self-assessments can help identify gaps in your skills or knowledge, guiding your learning and development efforts. This continuous cycle of feedback and self-reflection is key to advancing in your career.

Balancing Specialization and Diversification of Skills: Balancing specialization with diversification can be challenging but rewarding. Specializing in a specific area, like cybersecurity or cloud computing, can make you an expert in that niche. However, diversifying your skills will ensure you remain adaptable and can contribute in various roles. Consider your career goals and market demands to find the right balance. Being a specialist in one area while having a broad understanding of other IT domains can substantially enhance your career prospects.

Personal Development Plan: A personal development plan is a structured approach to achieving your career goals. It should outline the skills and knowledge you need to acquire, along with specific actions and timelines. Include short-term objectives leading towards your long-term career goals. Regularly review and update your plan to reflect your career progress and changes in the IT industry. A well-thought-out development plan keeps you focused and on track in your professional growth journey.

This chapter has journeyed through the multifaceted aspects of navigating the IT job market, from understanding its dynamics to mastering interviews and negotiating job offers. We've explored how continuous learning and adapting to emerging technologies are imperative for career advancement. This chapter serves as a guide to not only secure a desired position but also to grow and thrive in this field. Your career in Information Technology is a journey of continuous development, where adaptability and proactive learning are mandatory. As we transition into Chapter 6, we'll shift our focus to one of the most valuable skills in the industry. Networking is not just about making connections; it's about building meaningful relationships that can open doors to new opportunities and catalyze career growth.

Reflect and Apply Questions

1. **Market Trends:** Reflect on current IT market trends. How would these trends influence your job search strategy?
2. **Role of Networking:** Think about your professional network. How could you leverage it to uncover hidden job opportunities?
3. **Interview Preparation:** Consider an upcoming IT job interview. How would you prepare for technical and behavioral questions?
4. **Evaluating Job Offers:** Imagine you receive a job offer. What factors would you consider important when evaluating this offer?
5. **Adapting to Change:** Reflect on the rapid changes in IT. How would you ensure your skills and knowledge remain relevant in the evolving landscape?

Chapter 6: Mastering the Art of Networking

In the interconnected world of Information Technology, the power of networking can't be understated. Chapter 6 explores the essence and importance of networking within the industry, highlighting how it goes beyond simple social interactions to include building professional relationships that drive career growth and knowledge exchange. We'll explore the numerous benefits of networking, from uncovering hidden job opportunities to gaining invaluable industry insights and mentorship.

The Importance of Networking

In the IT industry, networking goes beyond mere socializing; it involves building professional relationships that are mutually beneficial. It encompasses connecting with peers, industry leaders, and potential mentors both in person and online. Understanding this broader definition of networking will help IT professionals effectively engage and leverage these connections for career development and knowledge sharing.

Benefits of Networking: Networking can open doors to job opportunities that are often not advertised publicly. It can also provide access to industry insights and trends, and facilitate the exchange of ideas and solutions to technical challenges. Building a robust network can even lead to mentorship opportunities and partnerships, providing support and guidance throughout one's career.

Networking Challenges and Overcoming Them: Networking can be daunting, especially for those who are introverted or new to the field. To overcome these challenges, start by setting small, achievable goals, like attending one networking event per month or reaching out to one new contact per week. Online platforms can serve as a less intimidating starting point, allowing you to connect with professionals globally. Engage in IT forums or LinkedIn groups where you can contribute to discussions, which can help in building confidence. Remember, networking is a skill that gets better with practice. It's also helpful to prepare conversation starters or questions in advance, making interactions at events less stressful.

Additionally, finding a networking buddy or mentor can provide support and guidance in navigating these environments.

Networking Success Stories:

1. **Software Developer at Hackathon:** Alice, a software developer, attended a local hackathon event organized by a prominent tech company. During the event, she showcased her skills by contributing significantly to a challenging project, which caught the attention of the company's team leads. The team was impressed with her coding abilities and problem-solving approach. Through networking at the event, Alice discussed the latest tech trends with the company representatives and expressed her interest in innovative software solutions. A few weeks after the hackathon, Alice received an invitation from the company for a formal interview. Her performance at the hackathon and her enthusiasm for technology led to a job offer as a software engineer, marking a meaningful step forward in her career.
2. **Professional Growth via LinkedIn:** John, an IT support technician, actively used LinkedIn to connect with industry experts. He regularly engaged with content posted by leaders in the field, sharing insightful comments and writing posts about his experiences and learnings. His proactive approach caught the attention of a senior executive, who appreciated John's in-depth knowledge and perspective. This led to a mentorship where the executive provided John with valuable industry insights and guidance. Over time, with the mentor's encouragement and advice, John gained the confidence to apply for higher roles. His efforts paid off when he landed a role as a network administrator, attributing much of his success to the mentorship and connections he fostered on LinkedIn.

Professional Relationship Building

Let's dive into strategies for cultivating and nurturing professional relationships within the field. This section will encompass techniques for expanding your network, fostering mentorship connections, and sustaining meaningful interactions.

Building Your Professional Network

When identifying potential contacts, focus on individuals who are active and influential in your areas of interest. Research professional groups and associations specific to your field, as these can be goldmines for networking. Attend industry meetups and seminars to meet these contacts in person. Following up after these events with a personalized message can solidify the initial connection. Also, consider reaching out to former colleagues or classmates who are now in the IT field; they could potentially offer valuable insights and introductions.

Effective Use of Professional Platforms: On platforms like LinkedIn, regularly update your profile with relevant skills and accomplishments to attract the right connections. Engage with content posted by others in your field by commenting and sharing your perspectives. Joining specific LinkedIn groups related to your interests can also provide networking opportunities. Don't hesitate to reach out with a personalized connection request to professionals you admire or wish to learn from. Regular activity on these platforms keeps you visible and in the minds of your network.

Networking Within Your Organization: Begin by pinpointing key influencers and decision makers in your organization who align with your areas of interest. Actively engage in internal training, workshops, and knowledge sharing sessions, as these are prime opportunities for meaningful interactions. Additionally, consider volunteering for cross-functional projects or committees, which not only broadens your internal network but also showcases your skills and initiative to a wider audience. Building relationships with colleagues from various departments can offer diverse perspectives on the organization and reveal hidden career opportunities.

Importantly, maintain these connections consistently, even if you transition to a different role or department. Regularly sharing updates about your projects or career developments can keep these relationships alive and fruitful. Integrating these strategies into your daily professional life will enable you to cultivate a robust internal network, laying a foundation for long-term career growth and opportunities within your organization.

Expanding Your Network: Broadening your professional network beyond your immediate workplace or known contacts will help in gaining diverse perspectives and uncovering new opportunities. Seek out industry associations, clubs, or groups that resonate with your IT interests and actively participate in their activities. These associations are not just platforms for meeting new people; they are hubs for sharing knowledge and trends in your field.

Consider contributing to these groups by giving talks, presentations, or volunteering for leadership roles, which positions you as an active and knowledgeable member of your community. Such involvement not only expands your network but also enhances your professional reputation. Always be prepared to share your contact information, and follow up with new connections promptly after events. Keep in mind that expanding your network is about forging genuine connections based on shared interests and mutual respect, rather than merely increasing your contacts list. This approach can lead to unexpected collaborations and opportunities that might not be available within your immediate professional circle.

Real World Example: A junior developer attends a major tech conference, focusing on networking opportunities. They engage with industry leaders during panel discussions, making a positive impression with their knowledge and enthusiasm. At social events, they connect with a senior developer, sharing interests and career aspirations. This connection leads to a mentorship offer, providing invaluable industry insights. A company repressentative, impressed by their active participation and technical questions, later offers them a job.

Embracing Diversity and Inclusion in Your Network: In the realm of IT, where innovation and collaboration are pivotal, diversity and inclusion within your professional network are more than just ethical imperatives; they are strategic assets. A diverse network brings together varied perspectives, experiences, and skills, enriching your understanding of global IT trends and challenges. By connecting with professionals from different cultures, backgrounds, and areas of expertise, you gain insights into a wide array of practices, regulations, and problem-solving approaches. This diversity fosters creativity and innovation and also equips you to navigate the complexities of international laws and workplace environments effectively. The IT industry is global, and having a network that reflects this diversity can provide you with an invaluable edge. Additionally, inclusive networking practices demonstrate your commitment to creating a professional environment where all voices are heard and valued. By consciously building and fostering a diverse and inclusive network, you contribute to a more equitable industry and enhance your own career prospects, opening doors to opportunities across the world.

Maintaining Professional Relationships: Schedule regular check-ins or catch-up meetings with your contacts. Offering help or advice when your contacts are in need can strengthen these relationships. Celebrate their achievements and milestones, as this shows you value the relationship beyond personal gain. Strong professional relationships are built on mutual respect and genuine interactions.

Leveraging Alumni and Academic Networks: Reach out to your alma mater's career services, which often have networking events and resources for alumni. Attend alumni gatherings and participate actively. Connect with former professors or faculty members who can provide industry insights and connections. Join online alumni groups where job opportunities and industry news are often shared. Your educational network is a lifelong resource that can provide valuable connections throughout your career.

The Power of Casual Conversations: Always be open to engaging in conversations in various settings, as you never know where a valuable connection might be made. Share your IT interests and experiences in these casual settings without being overbearing. Be genuinely interested in what others do – often, this interest is reciprocated, leading to discussions about professional opportunities. Listen actively and look for cues that could indicate a mutual professional interest. Carry a business card or be ready to connect on social platforms to continue the conversation later. Networking opportunities are all around us in everyday life.

Effective Networking Techniques

In the world of technology, understanding how to effectively conduct self-research is paramount for networking success. Online platforms are a goldmine for networking research. Here's how to make the most of them:

- **LinkedIn:** Use LinkedIn to follow industry leaders, join relevant IT groups, and engage with content in your field. Customize your feed to get updates from key influencers and companies.
- **Social Media Channels:** Follow tech thought leaders on platforms like X and YouTube. Engage with their content and stay updated with industry discussions and trends.
- **Meetup and Event Platforms:** Platforms like Meetup.com and Eventbrite are excellent for finding IT networking

events, webinars, and workshops. Tailor your search to your specific interests and set alerts for upcoming events.

- **Google Alerts and RSS Feeds:** Set up Google Alerts for specific IT topics, technologies, or influencers to receive timely updates. Use RSS feed aggregators to consolidate news and articles from your favorite IT blogs and websites.

Preparing for Networking Events: Before attending networking events, set clear objectives for what you want to achieve. Research the event, including the type of professionals attending, and plan whom you would like to meet. Prepare a brief introduction about yourself that briefly outlines your background and areas of interest. Think about questions you can ask to spark meaningful conversations. Dress appropriately for the event and bring business cards. Preparation can boost your confidence and make your networking experience more productive.

Effective Communication Skills: Practice active listening, showing genuine interest in what others have to say. Be clear and concise when talking about your professional background and aspirations. Use storytelling to make your experiences more engaging and memorable. Avoid dominating the conversation; networking is about mutual exchange. Good communication can leave a lasting positive impression on new contacts.

Creating a Lasting Impression: Be authentic and personable in your interactions. Share your passion and enthusiasm for your work and interests. Focus on creating a meaningful connection rather than just exchanging business cards. Be memorable by sharing unique insights or experiences relevant to your field. Follow up with a personalized message referencing specific parts of your conversation. A lasting impression often leads to lasting professional relationships.

Online Networking Etiquette: When reaching out to new contacts, start with a personalized message that mentions why you're interested in connecting. Participate actively in online discussions and groups, offering valuable insights. Maintain professionalism in all your online interactions. Regularly update your profile to reflect your current professional status. Online etiquette plays a big role in building and sustaining a quality image.

Follow-Up: This is important in networking. After meeting new contacts, send a personalized email or LinkedIn message expressing your appreciation for the conversation. Mention specific points from your discussion to reinforce the connection. Invite them for a coffee meeting or a more formal informational interview if appropriate. Regular follow-ups help transform initial meetings into lasting professional relationships.

Leveraging Networking for Career Opportunities: Use networking strategically to discover job opportunities. Express your career interests clearly when conversing with new contacts. Ask about openings in their companies or industry insights. Offer your skills or collaboration in areas of mutual interest. Networking isn't just about finding a job; it's about building relationships that might lead to opportunities in the future.

Going Beyond Your Comfort Zone: Networking beyond your immediate circle can be daunting but rewarding. Challenge yourself to attend events or join groups outside your usual network. Engage with experts from different IT specializations or industries. These interactions can provide fresh perspectives and unexpected opportunities. Stepping out of your comfort zone can greatly expand your network and open new doors in your career.

Utilizing social media for Networking

Select platforms that align best with your professional goals. LinkedIn is ideal for formal networking and career-focused content, while X (Twitter) offers more dynamic interactions with IT professionals and thought leaders. Choose platforms where your target network is most active, and where you can comfortably engage.

Creating a Strong Professional Profile: Your social media profiles should effectively showcase your IT expertise and achievements. Ensure your profiles are complete, up-to-date, and reflect your current career goals. Use a professional photo and a bio that clearly articulates your skills and areas of interest in IT.

Sharing and Publishing Content: Share and create content that demonstrates your expertise and interest. Regular posts about industry trends, insights, or personal experiences help establish your professional brand. Quality content can attract connections and open doors to new opportunities.

Social Media Etiquette and Professionalism: Maintain a professional demeanor in all of your online interactions. Be respectful, constructive, and positive in comments and posts. Avoid controversial or polarizing topics; focus on building an image that resonates with your IT career aspirations.

Measuring the Impact of Social Media Networking: Evaluate your social media networking effectiveness by monitoring your growing connections and engagement levels. Assess if your interactions lead to professional opportunities or valuable insights. Adjust your strategy based on these observations to optimize your social media networking efforts.

Mentorship and Its Role in Networking

Mentorship in Information Technology goes beyond basic networking; it's a relationship where experienced professionals provide guidance to those newer in the field. A mentor can offer advice, share experiences, and help navigate the complexities of the industry. This relationship is beneficial for gaining deeper insights into career paths, overcoming technical challenges, and understanding industry dynamics. A good mentor not only imparts knowledge but also helps in developing critical thinking and problem-solving skills. Mentorship is about personal and professional growth, facilitated through the wisdom and experience of someone who has navigated similar paths.

Finding the Right Mentor: Identifying the right mentor involves understanding your own career goals and seeking someone whose experience aligns with these objectives. Networking events, professional associations, and LinkedIn are great places to start looking for potential mentors. It's important to choose someone who not only has the expertise but also the willingness and capacity to mentor. A good mentor should be someone you respect professionally and feel comfortable communicating with openly. Approach potential mentors with a clear idea of what you seek from the mentorship and be respectful of their time and commitments.

Building a Productive Relationship: A successful mentor-mentee relationship is built on mutual respect, trust, and clear communication. It's important to set goals and expectations at the outset. Regular meetings can help keep the relationship focused and productive. As a mentee, be open to feedback and willing to act on the advice given. Remember, this relationship is a two-way street; your engagement and enthusiasm can also provide fresh perspectives and satisfaction to your mentor.

Expanding Your Mentorship Network: While traditionally, mentorship may evoke the image of a single, sage advisor, modern career trajectories often benefit from a diverse array of mentors. These mentors can encompass various facets of experts and personal development, including technical expertise, soft skills refinement, and industry specific insights. For example, aspiring IT pros may seek guidance from a network mentor well-versed in the intricacies of infrastructure and connectivity, a soft skills mentor to refine communication and leadership abilities, and a cybersecurity mentor to navigate the complexities of information security. By cultivating a multifaceted mentorship network, in-dividuals can gain holistic support tailored to their unique aspirations and challenges, fostering comprehensive growth and success in the dynamic IT landscape.

Giving Back - Becoming a Mentor: As you progress in your career, consider becoming a mentor to give back to the community. Mentoring others can enhance your own skills, broaden your perspective, and strengthen your leadership abilities. It's a rewarding way to contribute to the growth of upcoming professionals and to the industry as a whole. Being a mentor also expands your network and can offer new opportunities.

Navigating Challenges in Mentorship: Like any relationship, mentorship can face challenges such as misaligned expectations or communication issues. It's important to address any concerns early and openly. Set realistic expectations and be flexible in the relationship. If a mentorship isn't working out, it's okay to respectfully end the arrangement and seek a better fit. The goal is mutual growth and development, so finding the right mentor-mentee match is key.

Nurturing Professional Relationships

Transforming initial networking contacts into long-term professional relationships requires consistent effort and genuine interest. Focus on understanding the needs and interests of your contacts and look for ways to support them. Engage in meaningful interactions, share relevant information, and offer assistance when possible. Regular, non-intrusive communication helps in building trust and rapport over time. Strong professional relationships are built on a foundation of mutual respect and shared interests.

Regular Communication and Updates: Be sure to keep your network informed about your developments. Regular updates, whether through social media, emails, or personal messages, keep your contacts engaged and aware of your progress. It's about sharing successes, seeking advice, and even discussing industry trends. These updates maintain your presence in their minds, increasing the likelihood of them thinking of you for opportunities. Always aim for a balance in communication frequency – enough to stay connected but not so much that it becomes overwhelming.

Providing Value to Your Network: To maintain robust professional relationships, focus on providing value to your contacts. This could be in the form of sharing insightful articles, offering your expertise, or connecting them with other experts. Demonstrating that you're a resourceful and supportive contact encourages a reciprocal attitude. Make these relationships beneficial for all involved, fostering a sense of community and collaboration.

Recognizing and Celebrating Achievements: Congratulate your connections on their successes, whether it's a promotion, a successful project, or a personal milestone. This not only shows that you value the relationship but also fosters a positive and supportive network environment.

Reconnecting with Old Contacts: Re-establishing connections with former colleagues or acquaintances can be beneficial. Reach out with a personalized message, recalling your previous interactions and expressing interest in their current endeavors. Reconnecting can open doors to new opportunities and insights, as these contacts may have ventured into different areas of IT or advanced in their careers.

Assessing Networking Impact

Tracking Networking Outcomes: Keep a record of networking outcomes. Maintain a log of contacts made, conversations had, and any follow-up actions taken. Note any direct opportunities that arose from networking efforts, such as job leads, mentorship connections, or collaborative projects. This record not only demonstrates the tangible results of your networking but also helps you understand which activities are most beneficial for your career growth.

Analyzing Networking ROI: To evaluate the return on investment of networking, consider both the tangible and intangible benefits. Quantify the time and resources spent on networking against the opportunities and knowledge gained. Reflect on how networking has influenced your career trajectory. This analysis can help you focus your future networking efforts where they are most effective.

Setting and Reviewing Goals: Establish specific, measurable goals for your networking activities, such as expanding your network by a certain number or gaining insights into a particular IT specialty. Regularly review these goals to assess your progress and refine your approach. Setting goals turns networking from a passive activity into a strategic tool for career development.

Reflective Practices: Incorporating reflective practices into your routine can enhance the qualitative understanding of networking's impact on your career. Set aside time to reflect on the interactions you've had and the knowledge you've gained from your network. Assess how these interactions contribute to your professional development and career satisfaction. Reflective practices can offer insights into the more subtle, yet equally important, benefits of networking, such as increased confidence and a broader understanding of the IT industry.

As we conclude Chapter 6, remember that mastering the art of networking is a continuous journey, one that requires persistence, adaptability, and genuine engagement. The strategies and insights shared in this chapter are designed to empower you in building a diverse and robust professional network, essential in navigating the dynamic landscape of IT. Networking goes beyond mere connection-building; it is about creating meaningful relationships that foster mutual growth and learning. By embracing diversity and inclusion, actively engaging in various networking platforms, and extending your network beyond familiar circles, you position yourself to access a wealth of knowledge, opportunities, and support. The art of networking, when practiced diligently, can open doors to unexpected and rewarding career paths, enhancing your professional life in ways you may not have imagined. As you step forward, carry these networking principles with you, integrating them into every aspect of your journey.

Now, as we turn the page to Chapter 7, we will probe into the nuances of acing interviews, from technical to behavioral aspects, and explore effective strategies for climbing the career ladder. This chapter aims to equip you with the tools and knowledge necessary to succeed in interviews and make strategic moves that propel your career forward.

Reflect and Apply Questions

1. **Networking Strengths and Weaknesses**: Reflecting on your current networking strategies, identify one area where you feel most confident and one area where you see room for improvement. How can you leverage your strengths to address these improvements?

2. **Maximizing Networking Opportunities:** Think about a recent networking opportunity you had (or missed). What could you have done differently to make the most of that opportunity? How will you apply that lesson to future networking events?
3. **Optimizing Online Presence:** Considering your online presence, particularly on professional networking platforms like LinkedIn, how effectively do you think it represents your professional identity? What steps can you take to enhance your online networking profile?
4. **Analyzing Successful Interactions:** Recall a successful networking interaction you've had in the past. What made it successful? How can you apply these elements to future networking situations?
5. **Implementing Networking Strategies:** How do you plan to integrate the networking techniques discussed in this chapter into your ongoing professional development? Set one specific networking goal for the next month.

Chapter 7: Mastering Interviews and Career Advancement in IT

In the innovative sphere of Information Technology, excelling in interviews and strategically navigating your career path are as important as your technical skills. This chapter digs into the distinctions of interviews, offering insights into preparing for both technical and behavioral questions, and understanding the unique dynamics of interviews in the tech sector. We'll explore effective strategies for showcasing your skills and aligning with company culture, and provide guidance on how to handle various interview formats. Additionally, this chapter addresses the importance of continuous learning and professional development, guiding you on how to identify growth opportunities and strategically network for career progression. As you journey through the cutting-edge realms of IT, mastering these elements is key to unlocking doors to new opportunities and thriving in your chosen career path.

Interview Preparation

Let's discuss essential aspects of preparing for IT interviews, covering both technical and behavioral components. Then we will look at strategies for post-interview follow-up.

Understanding the IT Interview Landscape

In the IT industry, interviews can broadly be classified into technical and behavioral categories. Technical interviews focus on assessing your hard skills, like coding and system design, often featuring real-time problem-solving tasks. Behavioral interviews, on the other hand, aim to understand your soft skills, such as teamwork and decision-making. Additionally, some companies conduct case studies or group interviews to evaluate your analytical and collaborative abilities in a more dynamic setting. It's necessary to be prepared for each type, as they collectively provide a comprehensive view of your fit for the role and the organization.

Understanding a company's culture and values is beneficial in an interview. Companies often seek candidates who not only have technical prowess but also align with their cultural ethos. During the interview, you might encounter questions designed to gauge how well you fit into their work environment. Researching the company's values, mission, and work culture beforehand can help you tailor your responses to resonate with their expectations. Demonstrating an understanding of and compatibility with the company culture can substantially elevate your candidacy.

Interview Formats: IT interviews come in various formats, including in-person, remote, and panel interviews. In-person interviews offer direct interaction with the team, while remote interviews, increasingly common in today's global workplace, require comfort with digital communication tools. Panel interviews, involving multiple interviewers, can be challenging as they assess how you handle diverse perspectives and questions. Each format demands a different set of preparation strategies, from mastering virtual communication tools to engaging effectively with multiple interviewers.

Navigating a Virtual Interview: Mastering virtual interviews has become an indispensable skill for IT professionals. Virtual interviews, often conducted via video conferencing platforms, require a unique set of considerations compared to traditional in-person interviews. It's important to ensure a professional setting, with minimal background distractions and good lighting. Equally vital is testing your technology beforehand to avoid technical glitches. Additionally, maintaining eye contact with the camera, rather than the screen, helps in creating a more personal connection. It's also advisable to have a backup plan, such as a phone number to call, in case of technical issues. Preparing for a virtual interview also means anticipating and practicing responses to remote work-related questions, demonstrating your ability to work effectively in a virtual team. Overall, the key to excelling in a virtual interview lies in preparation, technical readiness, and effective communication skills.

Evaluating the Interviewer: An effective interview strategy involves understanding your interviewer's perspective. Different interviewers have unique styles, from technical experts focusing on skills to HR representatives assessing cultural fit. Adapting your responses to suit the interviewer's style can set you apart from other applicants. This might mean providing more technical detail for a tech lead or focusing on teamwork and collaboration for an HR manager. Learning to read the interviewer's cues and adapting accordingly is a valuable skill in navigating the interview successfully.

Common Interviewing Methods: The IT sector employs a variety of interviewing methods. Whiteboard coding and online assessments are popular for evaluating technical skills, where you're asked to solve coding problems in real-time. Portfolio reviews are common for design and development roles, where your past work is critically evaluated. These methods test not just your technical ability but also your problem-solving approach and creativity. Understanding these methods and practicing accordingly can greatly improve your performance in these interviews.

The Global Scene of Interviews: In the global IT industry, interviewing styles can vary significantly across regions. While some cultures may emphasize technical expertise, others may value a more holistic view encompassing soft skills and personality traits. This diversity in interviewing approaches requires a flexible and well-informed preparation strategy. Familiarizing yourself with the interviewing norms and expectations of different regions and companies is beneficial, especially for roles in multinational corporations or if you're considering working abroad. Cultural sensitivity and adaptability can be your assets in such diverse interview settings, enabling you to navigate these differences effectively and showcase your skills in a globally competent manner.

Technical Interview Preparation

Start by understanding the structure of technical interview questions. These questions typically revolve around algorithms, data structures, and system design, each testing different aspects of your IT skills. Common themes in these questions include problem-solving, logical thinking, and efficiency. Recognizing these patterns can help you prepare more effectively, as it's not just about knowing the right answer, but also about demonstrating a clear and efficient problem-solving approach.

Researching Tips and Tricks: Gleaning insights from those who have been on either side of the interview table is invaluable. Experienced interviewers and candidates often share core insights about popular questions and interview structures specific to IT roles. Researching forums, blogs, and professional networks where such individuals share their experiences can provide a wealth of knowledge. It's not just about understanding the types of questions asked, but also about grasping the underlying intent and preferred responses. This insight helps tailor your preparation, focusing on areas that are frequently emphasized and understanding the rationale behind different question types.

Post-Interview Reflection and Note-Taking: Whether the interview was successful or not, analyzing your performance can offer substantial growth opportunities. Jot down the questions you remember, your responses, and areas you felt less confident in. This post-interview review allows you to identify knowledge gaps and refine your approach for future interviews. Every interview is a learning experience – understanding what went well and what didn't will benefit continuous improvement. Viewing each interview as a stepping stone, rather than a one-off event, can transform your approach and effectiveness in future opportunities.

Mock Interviews and Real-Time Problem-Solving: Mock interviews are a vital part of preparing for technical interviews. They simulate the interview environment, allowing you to practice thinking on your feet and verbalizing your thought process. Real-time problem-solving during these mock sessions can highlight areas needing improvement, helping you to develop a clear and concise approach to articulating your solutions. Seek a mentor or fellow professionals to help with a mock interview.

Handling Technical Interview Pressure: Technical interviews can be high pressure situations. Developing strategies to manage stress, such as deep breathing or positive visualization, can be helpful. Practicing under timed conditions can also help build confidence. Staying calm allows for clearer thinking and a more focused approach to solving complex problems. Ask questions for more detail on the situation. This could help connect the dots to the answer. Most of all, be honest. If you get questions where you are unsure of the answer, try your best but don't be ashamed to say you don't know. In a scenario like this, make sure to be enthusiastic about your willingness to learn. Techniques and concepts can be taught, but a desire to learn can't.

Feedback and Continuous Improvement: Feedback, whether from mock interviews or actual ones, is essential for growth. Constructive criticism helps identify areas of strength and those needing improvement. Embrace this feedback and use it to refine your techniques and understanding continuously. There is nothing wrong with asking what the interviewer feels you could improve on. Viewing the interviewer's feedback as constructive can provide an invaluable learning opportunity. They may highlight a skill gap that had previously gone unnoticed, yet is critical for the position. By addressing this area for improvement, you will be better equipped to excel in future interviews.

Behavioral Interview Mastery

Behavioral questions in IT interviews are designed to assess how you've handled past situations professionally. They provide insight into your problem-solving, teamwork, and adaptability skills. Employers use these questions to gauge how you might perform in their workplace. To prepare, focus on core competencies like leadership, time management, and conflict resolution. Understanding the intent behind these questions will enable you to tailor your responses to demonstrate how your experiences align with the role's requirements.

Storytelling Techniques for Behavioral Responses: The STAR technique is an effective strategy for answering behavioral interview questions. This approach structures your response by detailing a specific **Situation**, outlining the **Task** required, describing the **Actions** you took, and concluding with the **Result** achieved. When applying this method, it's essential to strike a balance between demonstrating your technical abilities and soft skills. For instance, you could narrate an instance of a complex project (Situation), your specific responsibilities (Task), the particular steps you executed (Action), and the successful outcome, perhaps emphasizing collaboration or innovative problem-solving (Result).

Demonstrating Soft Skills Effectively: In IT interviews, demonstrating soft skills such as effective communication, collaboration, and creativity is vital. Adapt your examples to illustrate these skills in action. For instance, discuss a time when your communication skills led to a successful project outcome or how your creative problem-solving resolved a technical issue. These examples should be relevant to the role you're applying for, showing how your soft skills complement your technical abilities.

Preparing for Common Behavioral Questions: Prepare for common behavioral questions like 'Describe a time when you faced a challenging situation at work.' Research and list frequently asked questions, then prepare responses that reflect your experiences. Authenticity is key; rehearse your stories to ensure they sound natural and reflect your true professional experiences. This preparation will make you more confident and articulate during the actual interview.

Post-Interview Reflection: Reflect on your behavioral responses after the interview to identify strengths and areas for improvement. Think about the clarity, relevance, and impact of your answers. Did you effectively demonstrate the desired competencies?

Were your stories engaging? This reflection will benefit your continuous development, allowing you to refine your storytelling and interpersonal skills for future interviews.

Interview Follow-up

After your interview, a well-crafted follow-up can make a significant difference. Send a thank-you email within 24 hours, expressing your gratitude for the opportunity and reiterating your interest in the position. This message should be concise, personalized, and reflect a key point from your interview to show engagement. Avoid sounding overly persistent or pushy in your communication. If you haven't heard back within the timeline provided, it's appropriate to send a polite inquiry. This follow-up demonstrates your enthusiasm and professionalism, keeping you in the minds of the interviewers.

Handling Rejections: Receiving a rejection can be disheartening, but it's a chance for growth. Respond graciously to the rejection, thanking the interviewers for their time and consideration. Express your continued interest in the company and inquire about future opportunities. This maintains a positive relationship with the potential employer. Reflect on the interview experience to identify areas for improvement. If comfortable, ask for feedback on your performance. This information is invaluable for enhancing your skills and interview techniques for future opportunities.

Staying Organized and Following Through: Staying organized in your job search is important, especially when managing multiple interviews. Keep a detailed record of each interview, including dates, contact information, and notes about the conversation. This organization aids in timely follow-ups and keeps you prepared for any further communication. It also helps in assessing and comparing different job opportunities, ensuring you make well-informed decisions.

Building Long-Term Connections: Interviews are not just about landing a job; they're opportunities to expand your professional network. Even if you don't get the job, maintaining a connection with the interviewer can be beneficial in the long run. Connect with them on professional networks like LinkedIn, and keep the conversation going by sharing industry insights or articles. These connections can lead to future opportunities or valuable professional relationships. Every interaction can be a step towards future opportunities, so treat each interview as a valuable connection in your career journey. This approach can open doors you might not have anticipated and foster a rich, diverse professional network that supports your career growth over time.

Advancing Your IT Career

Next, we explore strategies and tips for career advancement in the industry, including continual learning, skill development, and navigating transitions within the field. Finally, we will cover personal development for IT pros.

Career Advancement Strategies

Identifying growth opportunities is pivotal to career advancement. Stay abreast of emerging technologies and industry trends, as they often signal new roles and specializations. Engage in continuous learning and be open to taking on challenging projects that stretch your abilities. Attend industry conferences and webinars, which can be excellent sources of information on future trends. Building a career in IT is about staying adaptable and always being ready to embrace change, ensuring your skills and knowledge remain relevant.

Strategic Networking: Networking is the backbone of career advancement within IT. It's not just about meeting new people; it's about building meaningful relationships with peers, mentors, and industry leaders. Engage actively in professional groups and online communities. Seek mentorship opportunities, as mentors can provide guidance, support, and exposure to new opportunities. Effective networking is a reciprocal process; be ready to offer help and insights to your connections as well.

Skill Development and Continuous Learning: The rapid pace of technological change in IT requires a commitment to continuous learning. Upskilling is indispensable to stay competitive. Identify areas where your skills need enhancement and pursue relevant certifications or training programs. Online courses, workshops, and industry certifications can provide the knowledge and credentials needed to advance. Balancing work with learning can be challenging but is necessary for long-term career growth.

Navigating Career Transitions: Transitioning to new roles or domains within IT requires careful planning and execution. Whether you're moving up the career ladder or shifting to a new specialization, assess the skills and knowledge required for the new role. Network with professionals already in the desired role for insights and advice. Be prepared for a learning curve and remain flexible as you adapt to new responsibilities. Successfully navigating career transitions often involves embracing new challenges and stepping out of your comfort zone.

Work-Life Balance and Managing Burnout: In a demanding field like IT, it is important to maintain a healthy work-life balance. Set clear boundaries between work and personal time. Engage in activities outside of work that rejuvenate you, such as hobbies or exercise. Be aware of the signs of burnout, like chronic fatigue or diminished interest in work. Prioritizing self-care and seeking support when needed will make for a sustainable and fulfilling career.

Impact of Industry Trends and Market Dynamics: To ensure career longevity, stay informed about industry trends and market dynamics. Technological advancements and economic shifts can rapidly change the landscape, affecting job roles and opportunities. Keep an eye on industry reports and market analyses to anticipate changes. Being proactive and adapting your career strategy in response to these changes can position you for continued success in the industry.

Effective Goal Setting and Career Planning: For sustained success in the IT field, get into the habit of effective goal setting and career planning. Begin by defining clear, realistic career goals that align with your interests and strengths. Break these goals into achievable milestones, creating a roadmap for your career progression. Perform regular self-assessments to monitor your progress and make adjustments as needed. Attend industry events and engage in professional development activities to stay aligned with your goals. A well-thought-out career plan is a dynamic document; it should evolve as you grow professionally and as the industry changes.

Continual Learning and Skill Development

Identifying your skill gaps is the first step towards effective learning and development. Regular self-assessment helps in pinpointing areas where you need improvement. Use feedback from supervisors, peer reviews, and self-reflection to identify these gaps. Consider both technical skills and soft skills, as both are required in the IT industry. Once identified, these gaps guide your learning journey, helping you focus on acquiring skills that enhance your professional value.

Setting Learning Goals: Set clear, achievable learning goals to ensure skill development. These goals should align with your career aspirations and address identified skill gaps. Be specific about what you want to achieve and set a realistic timeline. Goals can range from mastering a new programming language to improving project management skills. Regularly revisit and adjust these goals as your career progresses and as industry demands evolve.

Choosing the Right Resources: With an abundance of options, from online courses to traditional workshops, evaluate resources based on their relevance to your goals and learning style. Seek recommendations from peers or mentors and consider the credibility of the source. Also, look for resources that offer practical, hands-on experience, as it is valuable in the IT field.

Learning Methods for IT Professionals: Online courses offer flexibility, while workshops and conferences provide hands-on experiences and networking opportunities. Peer learning groups and forums can be invaluable for collaborative learning and problem-solving. Experiment with different methods to find what works best for you, keeping in mind that a combination of approaches is often the most effective.

Adapting to Technological Changes: In the IT industry, adapting to technological changes is a frequent occurrence. Stay curious and open to learning new technologies as they emerge. Participate in training sessions and certifications that focus on upcoming technologies. Being adaptable ensures your skills remain relevant and positions you as a valuable asset to your organization. The ability to quickly learn and apply new technologies is a cherished trait of successful IT experts.

Navigating Career Transitions

Career transitions, whether moving to a new role or a different company, often involve a mix of excitement and apprehension. It's important to recognize the emotional phases, from the initial excitement to possible uncertainty. Acknowledge these feelings as a natural part of the process. Reflect on your career path and what prompted the transition. This self-awareness can be instrumental in navigating the change smoothly and aligning your career with your long-term goals.

Evaluating New Opportunities with a Global Perspective: The IT industry is increasingly interconnected, and opportunities in one part of the world can be influenced by trends and developments in another. For instance, a growing demand for cloud computing expertise in Europe might offer opportunities for individuals specializing in this area, even if they are based on a different continent. When evaluating new roles, consider not only the local market conditions but also global industry trends. Assess how emerging technologies are being adopted in various regions and how that impacts the demand for specific skills. Additionally, understanding the work culture, professional practices, and regulatory environment of different countries can provide a more comprehensive view of the global landscape. This broader perspective will enable you to make more informed decisions about your career path, opening up possibilities for international collaborations or even relocation.

Real-World Example: Consider the story of Allen, an IT professional faced with two job offers. The first offer was from a well-known, large tech company. The role came with a lucrative salary and the prestige of being associated with a major player in the industry. However, the competition for career advancement in such a large company was known to be intense.

The second offer came from a smaller, emerging tech company. While the salary was lower than the first offer, the company was in urgent need of skilled experts. The role promised more hands-on responsibilities and a significant opportunity for growth. After careful consideration, Allen chose the second offer. This decision was based on the alignment of the role with Allen's skills and the potential for rapid career advancement.

Allen's choice proved to be a turning point. In the smaller company, Allen was able to stand out by making impactful contributions to key projects. This visibility led to quicker promotions and, within a few years, Allen not only surpassed the salary level of the first offer but also gained a reputation as a vital player in the company's growth. Allen's story exemplifies the importance of assessing long-term career growth and personal skill alignment when evaluating new opportunities, rather than focusing solely on immediate financial gains.

Skill Mapping for Transition: Start by listing your current skills and compare them with the requirements of the new position. This process helps in identifying both transferable skills and areas where you might need additional training. Consider more than just technical skills such as soft skills like leadership or communication. This exercise prepares you for the new role and helps in highlighting your strengths during the interview process. Continuous skill development positively benefits successful career transitions in the IT industry.

Skill Mapping Exercise:

1. **Current Skills Assessment:**
   a) Create a comprehensive list of your current skills, both technical and soft skills.
   b) Rate your proficiency in each skill on a scale (e.g., beginner, intermediate, advanced).
2. **Desired Role Analysis:**
   a) Research the skills and competencies required for your desired role.
   b) Create a list of these skills, using job descriptions or industry standards as references.
3. **Gap Analysis:**
   a) Compare your current skills with those required for the new role.
   b) Identify gaps where your skills need improvement or where you need new skills.
4. **Action Plan:**
   a) For each identified gap, develop an action plan. This could include specific courses, training, or experiences needed to acquire or improve these skills.
   b) Set achievable goals with a timeline for each action item.
5. **Progress Tracking:**
   a) Create a system to track your progress in developing these skills.
   b) Regularly review and adjust your plan as needed.

Personal Branding During Transition: Your brand should reflect your new professional direction and highlight relevant skills and experiences. Update your LinkedIn profile, resume, and personal website to highlight this shift. This might include rewriting your professional summary or highlighting different aspects of your experience. A well-managed personal brand can enhance your visibility and attractiveness to potential employers in your new chosen field.

Preparing for the Transition: This includes both mental preparations, like setting expectations for a new role, and practical steps like financial planning. If your new role requires relocation, start planning early to ensure a smooth transition. It's also important to update or acquire any necessary certifications or training for your new role. Proactive preparation can help mitigate the stress associated with career shifts and set you up for success in your new position.

Adapting with a Global Lens: Adapting to new work environments in IT goes beyond adjusting to a new company; it often involves navigating diverse cultural landscapes. In today's global industry, you might find yourself working with colleagues, clients, or stakeholders from different parts of the world. Understanding and respecting cultural differences is important to successful collaboration. For example, communication styles, decision making processes, and work-life balance priorities can vary substantially across cultures. Being culturally sensitive and adaptable not only enhances your ability to work effectively in diverse teams but also positions you as a valuable asset in a globalized workplace. Embrace opportunities for cross-cultural interactions, whether through international projects, remote teams, or expatriate assignments. These experiences enrich your professional skill set and broaden your understanding of the global industry.

Networking and Research: When navigating career transitions, utilize effective networking and thorough research. Start by reaching out to your existing network for insights into potential roles or companies. Attend industry events and engage with professionals in the field you're transitioning into. As for research, go beyond the company's website; look at news articles, industry reports, and employee reviews.

As a real-world example, consider a project manager aiming to transition into cybersecurity. They could start by following cybersecurity experts on social media, attending relevant webinars, and joining cybersecurity groups. Engaging in these activities provides valuable insights and helps in building connections with experts in the field, leading to potential mentorship or job opportunities.

Managing Change and Resilience: Career transitions can be challenging, and it's important to build resilience during this time. Develop strategies to manage stress, such as maintaining a healthy work-life balance and seeking support when needed. Stay flexible and open-minded, and embrace the learning opportunities that come with change. Developing coping strategies, such as mindfulness or seeking out professional advice, can help you navigate this period effectively. Remember, each career transition is a stepping stone in your journey, offering valuable experiences that contribute to your overall growth.

Personal Development

Developing Leadership Skills: In IT, having leadership skills will set you apart regardless of your position. These skills involve more than just managing a team; they're about inspiring and guiding colleagues towards achieving common goals. Start by taking initiative in projects, mentoring junior team members, and leading by example. Attend leadership workshops and seek advice from peers and supervisors to continually improve. Understanding different leadership styles and adapting them to your work environment can greatly enhance your effectiveness.

Effective Communication: It's important to convey complex technical concepts in a way that non-technical stakeholders can understand. Practice active listening, clear and concise speaking, and considerate writing. Engaging in cross-departmental projects can provide opportunities to refine these skills. Effective communication is key to teamwork, problem-solving, and project success.

Cultivating a Problem-Solving Mindset: Develop this mindset by embracing challenges as opportunities to learn and grow. Engage in activities that push your thinking, like hackathons or complex project work. Study how others approach problems and adapt their strategies. Continuous learning and staying curious about new technologies and methodologies can also enhance your problem-solving skills.

Emotional Intelligence (EI) in the Workplace: Emotional intelligence involves self-awareness, empathy, and regulation of emotions in professional settings. Developing EI can lead to better team collaboration and conflict resolution. Seek feedback on your interpersonal skills and be open to change. Engaging in workshops or training can also be beneficial in enhancing EI.

Time Management and Productivity: These skills involve prioritizing tasks, setting realistic deadlines, and avoiding distractions. Techniques like the Pomodoro Technique or time blocking can improve focus and productivity. It's also important to set aside time for breaks and self-care to prevent burnout. Utilizing tools like project management software can also aid in managing tasks efficiently.

Continuous Self-Assessment and Feedback: Regular self-assessment and seeking feedback are important for personal growth. Set aside time to reflect on your achievements and areas of improvement. Constructive feedback from colleagues and supervisors can provide valuable insights into your performance. Be proactive in seeking opportunities for feedback, especially after completing projects or milestones. Use this feedback to set goals for future development.

Adaptability and Flexibility: Adaptability in the IT industry is about staying flexible in your approach to work and problem-solving. This means being open to new ideas, methodologies, and changes in project direction. Cultivate a mindset that embraces change as a constant, and views challenges as opportunities to learn. Engaging in diverse projects and continuously updating your skills can help maintain this adaptability.

Innovation and Creative Thinking: Encourage yourself to think outside the box and explore new ways of approaching problems. Collaborate with diverse teams to gain different perspectives and ideas. Stay informed about the latest industry trends and technologies to inspire innovative solutions. Creative thinking often comes from combining existing ideas in novel ways, so keep learning and exploring.

As we conclude Chapter 7, it's clear that mastering interviews and excelling in career advancement involves more than just technical acumen. It encompasses a holistic approach towards personal development, embracing leadership, effective communication, adaptability, and continuous learning. These skills enhance your career prospects and contribute to your overall professional growth. The journey in Information Technology is one of continual evolution, where personal and professional development go hand in hand. By focusing on these areas, you equip yourself with the tools necessary for a successful and fulfilling career in the industry.

As we shift our focus to Chapter 8, we explore the critical aspect of cybersecurity in the IT world. This chapter will guide you through the fundamentals of information security, underscoring its importance in every aspect of IT and how you can integrate a security first approach into your professional skill set.

Reflect and Apply Questions

1. **Professional Challenge:** Reflect on a recent professional challenge. How did you approach it and what could you have done differently using the skills discussed in this chapter?
2. **Continuous Learning:** How do you plan to integrate continuous learning into your daily routine to stay updated with the latest IT trends?
3. **Addressing Skill Gap:** Identify a skill gap you currently have. What steps will you take to address it in the next three months?
4. **Communication Impact Analysis:** Think of a situation where effective communication significantly impacted a project or team dynamic. How did it change the outcome?
5. **Leadership Experience:** Reflect on a recent project or task in your career where you had to lead, either formally or informally. What leadership qualities did you exhibit, and how did they impact the project's outcome?
6. **Adaptability Learning Experience:** Describe a time when adaptability or flexibility was crucial in your career. What did you learn from that experience?

Chapter 8: Developing a Security Mindset

This chapter takes center stage as a pivotal section within this book, diving into the expansive realm of cybersecurity and its undeniable significance in the contemporary digital sphere. The importance of cybersecurity in today's technologically driven world cannot be overstated. To comprehensively address this vast subject matter, I've decided to split it into two distinct chapters.

In this chapter, we embark on a journey to equip you with a fundamental understanding of the evolving cyber threats that saturate our interconnected world. We'll explore the dynamic landscape of cybersecurity, shedding light on both the threats and the proven best practices to counter them. As we progress, you'll gain the knowledge and insights needed to navigate the digital domain with confidence, making informed decisions and contributing to a safer online environment for yourself and your digital peers.

Foundations of Cybersecurity

In this section, we explore the foundational principles of cybersecurity required for building a strong security posture. We will discuss the cyber threat landscape, fundamentals of network security, and secure communications. These introductory elements form the basis of effective cybersecurity practices in organizations.

Introduction to Cybersecurity

Cybersecurity is the practice of defending digital systems, networks, and data from an array of potential threats and vulnerabilities. It encompasses an evolving landscape of strategies, technologies, and best practices. At its core, cybersecurity is about safeguarding the digital realm, ensuring that information remains confidential, unaltered, and accessible to authorized users.

Within the realm of cybersecurity, specialists employ a multi-faceted approach that includes threat detection, risk assessment, vulnerability management, incident response, and the implementation of security measures. It's a dynamic field where experts constantly adapt to counteract the ever-evolving tactics of cybercriminals.

History and Evolution of Cyber Threats: The story of cybersecurity is a tale of continuous evolution, shaped by the relentless ingenuity of both malicious actors and the defenders who rise to meet them. Understanding the history and evolution of cyber threats is necessary to appreciate the complexity of today's digital battleground.

In the early stages of computer history, cyber threats were basic, often experiments with new technology. With the rise of personal computers and the internet, these threats became more sophisticated, leading to the development of damaging viruses and worms capable of self-replication and system disruption. As online connectivity expanded, so did the variety of cyber threats. This era saw the emergence of more deceptive forms of malware, such as Trojans, which masquerade as harmless software, and spyware aimed at covertly collecting user data. Additionally, phishing attacks grew common, utilizing well-crafted deceptive emails to trick individuals into disclosing personal information.

In recent years, the digital landscape has witnessed the rise of ransomware, a particularly insidious form of malware that encrypts a victim's files, demanding a ransom for their release. Data breaches have become headline news, exposing the personal information of millions. Cyber espionage and nation-state-sponsored attacks have further complicated the cybersecurity landscape, blurring the lines between criminal activity and geopolitical warfare. This historical journey through the evolution of cyber threats highlights the relentless innovation and adaptability of malicious actors. It underscores the critical need for a robust cybersecurity mindset in an era where the stakes have never been higher.

Basic Terminology and Concepts: In the world of cybersecurity, one should have a foundational understanding of basic terminology and concepts:

- **Encryption**: Encryption involves encoding data to make it indecipherable to unauthorized parties. It safeguards sensitive information during transmission or storage. This ensures it is very difficult to decipher from anyone who isn't authorized to view it.
- **Firewall**: A firewall acts as a digital barrier that monitors and controls incoming and outgoing network traffic. It serves as a gatekeeper to prevent unauthorized access and protect against cyber threats.
- **Vulnerability**: A vulnerability is a weakness or flaw in a system or software that can be exploited by cybercriminals.

Identifying and patching vulnerabilities is crucial to maintaining security.

- **Patch**: A patch is a software update that addresses known security vulnerabilities. Regularly applying patches is required for keeping systems secure. Think of software updates for smart devices. These are often created to patch vulnerabilities found by the company that created the device.
- **Virus:** A type of malware that replicates by attaching itself to other programs. Very similar to a contagious flu virus to humans.
- **Worm:** A self-replicating malware that spreads across networks without user interaction.
- **Endpoint Security:** The process of securing various endpoints on a network, often defined as end-user devices such as mobile devices, laptops, and desktop PCs.
- **VPN (Virtual Private Network):** A tool that creates a secure, encrypted connection over a less secure network, like the Internet. Commonly used by organizations to securely provide access to remote workers.

Familiarity with these concepts provides a solid foundation for navigating the intricate world of cybersecurity.

Types of Cyber Threats: In the world of cybersecurity, a multitude of threats lurk in the digital shadows, each with its unique capabilities and intentions. Understanding the various types of cyber threats will assist in building a strong defense against potential attacks.

1. **Malware**: Malware, short for malicious software, is a broad category encompassing viruses, worms, Trojans, and spyware. These software programs are designed to infiltrate and compromise computer systems. Viruses and worms replicate themselves, spreading to other systems, while Trojans disguise themselves as legitimate software to deceive users. Spyware stealthily collects sensitive information without the user's knowledge.
2. **Phishing**: Phishing attacks are deceptive attempts to trick individuals into revealing personal information such as login credentials or financial data. Typically carried out via email, phishing messages appear legitimate but contain malicious links or attachments that can lead to identity theft or

financial loss. Can also be used via other methods besides email. **Smishing (SMS Phishing)** or **Vishing (Voice Phishing)**

3. **Ransomware**: Ransomware is a particularly menacing form of malware that encrypts a victim's files or entire systems, demanding a ransom for decryption. Falling victim to ransomware can result in data loss and financial extortion.
4. **Data Breaches**: Data breaches involve unauthorized access to and theft of sensitive information, such as personal data, credit card details, or company secrets. Breaches can occur due to various vulnerabilities, including weak passwords or inadequate security measures.
5. **Cyber Espionage**: Nation-states and threat actors engage in cyber espionage to steal sensitive information, conduct surveillance, or gain a competitive advantage. These sophisticated attacks often involve advanced techniques and are not limited to traditional criminal motivations.
6. **Insider Threats**: Insider threats arise from individuals within an organization who misuse their access and privileges to compromise security. This can be intentional or unintentional and may involve employees, contractors, or business partners.
7. **Distributed Denial of Service (DDoS) Attacks**: DDoS attacks overwhelm a target's online services or network resources, rendering them unavailable. Attackers use a network of compromised devices to flood the target with traffic, disrupting normal operations. Think of this like a Black Friday deal when it goes live. There are so many people trying to get the deal the second it comes online that the site can lag or crash. This is the same concept as a malicious actor performing a DDoS attack.
8. **Zero-Day Exploits**: Zero-day exploits target vulnerabilities in software or hardware that are unknown to the vendor or have not yet been patched. Cybercriminals and state-sponsored actors often exploit these vulnerabilities for their advantage.
9. **Advanced Persistent Threat (APT)**: A sophisticated cyberattack where an unauthorized user gains access to a network and remains undetected for a long period. Typically launched by well-organized groups, APTs aim to steal data or surveil activities rather than cause immediate damage.

Understanding these various cyber threats equips individuals and organizations with the knowledge needed to implement effective cybersecurity measures and respond swiftly to potential attacks. By staying informed and vigilant, we can collectively work towards a more secure digital environment.

The Impact of Cybersecurity: The consequences of cybersecurity, or the lack thereof, ripple across both individuals and organizations. For individuals, cybersecurity safeguards personal information, financial assets, and online identities. It empowers individuals to browse the internet, shop online, and communicate without fear of identity theft, fraud, or unauthorized access to their accounts.

In the realm of organizations, cybersecurity plays an indispensable role in protecting sensitive data, maintaining customer trust, and preserving brand reputation. The fallout from a breach can encompass financial losses, legal liabilities, and severe damage to the organization's image. The impact extends beyond the digital world into the physical, affecting day-to-day operations and the trust of stakeholders. Recognizing this wide-reaching impact stresses the urgency of prioritizing cybersecurity at both individual and organizational levels.

The Role of Cybersecurity in Everyday Technology: Cybersecurity is intricately woven into the fabric of everyday technology, shaping how individuals interact with and rely on digital devices and services. It serves as the guardian of digital experiences across various domains:

- **Smartphones**: In the palm of your hand, cybersecurity safeguards personal data, privacy, and the security of mobile applications. It ensures secure communication and guards against mobile-specific threats.
- **Social-Media**: As users share their lives on social platforms, cybersecurity protects personal information, login credentials, and online identities from prying eyes and cyber threats. Privacy settings and security features empower users to control their online presence.
- **IoT Devices**: Cybersecurity safeguards the growing ecosystem of interconnected IoT devices, from smart home appliances to wearable gadgets. It prevents unauthorized access and ensures the security of personal data collected by these devices.

Understanding the persistent role of cybersecurity in these everyday technologies empowers individuals to make informed choices, adopt safe online practices, and appreciate the significance of cybersecurity professionals in securing the digital world.

Best Practices: In the digital landscape, cybersecurity best practices serve as the foundation for safeguarding your digital life. Whether you're a seasoned IT professional or just embarking on your journey, understanding these practices is paramount. Here, we introduce you to some fundamental principles that form the basis of a security-conscious mindset.

1. **Password Hygiene**: Strong, unique passwords are your first line of defense. Regularly update your passwords, use a combination of upper- and lower-case letters, numbers, and special characters, and avoid easily guessable information like birthdays or names. Using a password manager is also a great option to keep your passwords secure. Some of the popular ones are LastPass and Bit Warden.
2. **Multi-Factor Authentication (MFA)**: Enable MFA wherever possible. MFA utilizes an app code, or token, that is required after you input your password. This code updates frequently, commonly every minute. This adds an extra layer of security by requiring a second verification step in addition to your password.
3. **Software Updates**: Keep your operating system, software applications, and antivirus programs up to date. These updates often include security patches that address known vulnerabilities.
4. **Awareness and Education**: Stay informed about the latest cybersecurity threats and scams. Educate yourself and your family or colleagues about phishing attempts, suspicious emails, and safe online behavior.
5. **Backups**: Regularly back up your important data to secure and offline locations. In case of ransomware or data loss, having backups ensures you can recover your valuable information.
6. **Privacy Settings**: Review and configure privacy settings on social media platforms, applications, and devices to limit the amount of personal information you share publicly.

7. **Secure Wi-Fi**: Use strong, unique passwords for your Wi-Fi network and router. Regularly change default router login credentials to prevent unauthorized access.
8. **Antivirus and Antimalware**: Install reputable antivirus and antimalware software on your devices. Regularly scan for and remove threats to your system.
9. **Phishing Awareness**: Be cautious about unsolicited emails, especially those requesting sensitive information or urging immediate action. Verify the legitimacy of emails and their senders.
10. **Physical Security**: Protect your devices physically. Use locks, secure your laptop, and avoid leaving devices unattended in public spaces.
11. **Security Updates**: Stay informed about security updates and patches for your devices and applications. Apply them promptly to address vulnerabilities.
12. **Secure Browsing**: Use secure, HTTPS-enabled websites when sharing sensitive information or making online transactions.
13. **Safe Downloads**: Be cautious when downloading files or applications from untrusted sources. Stick to reputable app stores and websites.

By incorporating these best practices into your daily routine, you can significantly reduce the risk of falling victim to cyber threats and contribute to a safer digital environment for yourself and those around you.

Understanding the Cyber Threat Landscape

The cyber threat landscape includes a vast and ever-evolving terrain where malicious actors continuously seek to exploit vulnerabilities for various purposes. It encompasses a wide spectrum of threats, ranging from individual hackers seeking financial gain to nation-states engaging in cyber espionage and warfare. Understanding this landscape is essential for anyone looking to navigate the digital world safely. It's a dynamic battleground where defenders must remain vigilant, adapt to emerging threats, and employ proactive security measures to safeguard digital assets and information.

The Motivations Behind Cyber Threats: Cybercriminals and threat actors are driven by diverse motives, each dictating their approach and targets. Financial gain remains a primary motivation, with cybercriminals seeking to steal money or valuable data. Hacktivism involves politically or socially motivated individuals or groups using cyberattacks to advance their causes. State-sponsored actors engage in cyber espionage and cyber warfare for political, economic, or military advantage. Understanding these motivations helps anticipate and counteract cyber threats effectively.

Cyber Attack Vectors: Cyber threats manifest through various attack vectors, each exploiting specific vulnerabilities. Phishing attacks deceive users through emails or messages, often leading to data breaches. Malware, including viruses, Trojans, and ransomware, infiltrates systems to cause harm or extort money. DDoS attacks overwhelm online services, rendering them in-accessible. Social engineering manipulates individuals into revealing sensitive information. Familiarizing oneself with these attack vectors is important for recognizing and mitigating threats.

Real-World Example: Meet Jane, a retired teacher in her 70s who enjoys staying connected with her friends and family through email and social media. One day, she receives an email that appears to be from her bank, notifying her of a security issue with her account. Concerned, she opens the email, which urges her to click on a link to verify her account details immediately.

The email looks convincing, complete with the bank's logo and an urgent tone. Without thinking twice, Jane clicks the link and enters her account information as prompted. Unbeknownst to her, she has just fallen victim to a phishing attack. The email wasn't from her bank but from cybercriminals seeking to steal her login credentials and access her financial information.

Defending Against Phishing and Social Engineering

1. **Verify the Source:** Always double-check the sender's email address and domain. Legitimate organizations usually use official email addresses. If you receive an email claiming to be from your bank, independently verify the contact information from their official website or by calling their customer support.
2. **Beware of Urgency:** Cybercriminals often create a sense of urgency to pressure victims into taking immediate action.

Take a moment to pause and think before clicking on any links or providing sensitive information.

3. **Don't Share Personal Information:** It is recommended to never share personal or financial information via email or phone calls, especially if you didn't initiate the contact. Legitimate organizations won't ask for sensitive details in unsolicited messages.
4. **Enable Multi-Factor Authentication (MFA):** Wherever possible, enable MFA for your online accounts. Even if attackers obtain your password, MFA adds an extra layer of security by requiring a second authentication step.

Targets and Victims: The spectrum of cyber threats spans an array of potential targets, from individuals and small businesses to critical infrastructure and government institutions. Cybercriminals often target entities with valuable data, seeking financial gain or leveraging stolen information for extortion. Nation-states focus on espionage, intellectual property theft, and potential disruption of rival nations. The diversity of targets high-lights the pervasive nature of cyber threats.

Notable Cyber Attacks: Examining past cyber-attacks provides valuable insights into the evolving threat landscape. High profile data breaches, such as the Equifax breach or the Target breach, showcase the impact on individuals and organizations. Large-scale ransomware attacks, like WannaCry, emphasize the urgency of cybersecurity measures. Nation-state operations, like the Stuxnet worm, demonstrate the geopolitical implications of cyberattacks. Studying these cases helps grasp the breadth and consequences of cyber threats. For more details on these cases, see the citations in the Appendix.

Emerging Threats: The world of cyber threats is in constant flux, with emerging threats presenting new challenges. Zero-day vulnerabilities, unpatched weaknesses in software or hardware, are prime targets for exploitation. AI-driven attacks employ machine learning for enhanced precision. IoT-related risks arise as interconnected devices expand, potentially becoming entry points for attacks. Understanding these emerging threats is vital for staying ahead of the evolving threat landscape.

The Dark Web and Cybercriminal Underground: These communities serve as hidden ecosystems where illegal activities thrive. Here, cyber criminals buy and sell stolen data, hacking tools, and services. Understanding the workings of these underground network's sheds light on the tools and tactics used by threat actors. It also underscores the importance of robust cybersecurity practices to defend against such threats.

Attribution and Challenges: Attributing cyberattacks to specific actors or nation-states is riddled with difficulties, making it one of the most complex aspects of cybersecurity. Cyber attackers are known to utilize a variety of methods to conceal their tracks, such as using compromised networks in third countries, employing anonymizing services, and mimicking the techniques of other groups. This deliberate obfuscation complicates the process of tracing the origins of an attack back to its source.

Furthermore, the role of cyber operations in the geopolitical arena cannot be understated. Nations engage in cyber activities not only for espionage but also for influencing global politics, which adds layers of complexity to attribution. For instance, a cyberattack that appears to be the work of a nation-state might actually be conducted by non-state actors aligned with that nation's interests, blurring the lines of responsibility.

Recognizing the perpetrator behind a cyber incident requires more than just technical analysis. It involves a comprehensive approach that combines the rigor of digital forensics, the insights of cybersecurity intelligence, and an understanding of the geopolitical landscape. Analysts must sift through digital evidence, assess the tactics, techniques, and procedures (TTPs) used, and correlate this information with intelligence on potential adversaries' capabilities and motives.

However, even with a multidisciplinary approach, absolute certainty in attribution is rare. The digital domain's anonymity and the constant evolution of cyber tactics often leave room for ambiguity. Acknowledging these challenges is critical not only for accurately assessing threats but also for formulating appropriate responses. As such, while striving to pinpoint the sources of cyberattacks is essential, it is equally important to recognize the inherent limitations and uncertainties that accompany attribution efforts in the cybersecurity field.

Fundamentals of Network Security: Network security is an indispensable component of the tech landscape, ensuring the protection of digital assets and data during their transit across interconnected networks. This aspect of cybersecurity involves an intricate blend of strategies, practices, and technologies specifically devised to shield networks from a myriad of threats. These threats not only seek unauthorized access but also aim to disrupt, steal, or damage valuable data, thereby jeopardizing the integrity and reliability of network systems.

Network security encompasses several core components:

1. **Access Control**: Ensuring that only authorized users and devices can access the network and its resources. This involves mechanisms like authentication and authorization, which verify user identities and define their access levels.
2. **Firewalls and Intrusion Prevention Systems (IPS)**: Firewalls act as gatekeepers, controlling incoming and outgoing network traffic based on predetermined security rules, while IPS systems monitor network and system activities for malicious activities or policy violations.
3. **Virtual Private Networks (VPNs)**: VPNs create a secure, encrypted connection over a less secure network, such as the internet, enhancing privacy and ensuring secure remote access.
4. **Anti-malware Solutions**: These tools protect against viruses, worms, spyware, and other types of malware that can infiltrate networks and cause damage or data loss.
5. **Data Encryption**: Encrypting data in transit and at rest provides a critical layer of security, making sensitive information unreadable to unauthorized individuals.
6. **Endpoint Security**: Securing every endpoint connecting to the network, from personal computers to mobile devices, is vital to prevent exploits that target these devices.
7. **Regular Audits and Monitoring**: Continuous monitoring of network traffic and regular security audits are required for detecting anomalies, breaches, or unpatched vulnerabilities in real-time.

Implementing a comprehensive network security strategy involves deploying the right mix of technologies and fostering a culture of security awareness among users. The weakest element of any security posture is the users. An organization can have the most robust security defense set up, but one end user is all it takes to let an attacker in. Education on safe online practices and the potential risks of phishing attacks, social engineering, and other cyber threats is crucial. In the scene of cybersecurity, staying informed of the latest threats and advancements in technology is paramount for maintaining strong network security.

Types of Network Threats: These threats come in various forms, from malware that can infiltrate and disrupt networks to determined cybercriminals attempting unauthorized access. DDoS attacks can overwhelm network resources, causing disruptions, while insider threats may arise from employees or trusted entities with access to sensitive data. Recognizing the diversity of threats equips individuals and organizations to identify potential vulnerabilities and take proactive measures.

Encryption and Data Protection: Encryption is a foundational aspect of network security. It involves encoding data to render it unreadable to unauthorized individuals or entities. Protocols like SSL/TLS are commonly used to secure data during transmission, safeguarding sensitive information exchanged over the internet. Proper encryption also extends to data storage, ensuring that even if physical devices are compromised, the data remains inaccessible without the decryption keys.

Network Security Best Practices: Ensuring network security requires a proactive approach. Best practices involve regular software updates to patch vulnerabilities, creating strong, unique passwords for network access, and conducting periodic security audits to identify weaknesses. User education plays a critical role in promoting safe online practices and awareness of potential threats. The combined effort of proactive measures and user involvement is pivotal in maintaining network security.

Security Challenges in Wireless Networks: Wireless networks, while providing the convenience of mobility and ease of access, also open up a spectrum of security vulnerabilities that attackers can exploit. The inherent nature of wireless communication, broadcasting data through airwaves, makes it more susceptible to interception and unauthorized access compared to wired networks. Recognizing and understanding the common wireless attacks is vital for implementing effective security measures.

Common Wireless Attacks:

1. **Eavesdropping (Sniffing)**: This passive attack involves intercepting and monitoring wireless communications to capture sensitive information such as passwords, credit card numbers, and personal data. Without strong encryption, data

transmitted over wireless networks can be easily captured by attackers within range.

2. **Rogue Access Points**: Malicious actors set up unauthorized access points that mimic legitimate ones to deceive users into connecting to them. Once connected, the attacker can monitor all traffic passing through the rogue access point, capturing sensitive information or injecting malware into the network. These are common in public places, which is why it is imperative to utilize a VPN if you need to transmit important information or data.
3. **Evil Twin Attacks**: Similar to rogue access points, evil twin attacks involve creating a malicious Wi-Fi access point with the same SSID (Service Set Identifier) as a legitimate one. Users are tricked into connecting to the fraudulent network, allowing attackers to intercept data.
4. **Man-in-the-Middle (MitM) Attacks**: In this scenario, attackers insert themselves between the user and the legitimate network to intercept, alter, or reroute communications. This can lead to data theft, session hijacking, and other malicious activities.
5. **Wireless Denial of Service (DoS)**: Attackers flood a wireless network with excessive traffic or send specific frames that disrupt the network, rendering it unavailable to legitimate users. This can be used to distract from other malicious activities or as a form of extortion.
6. **Wi-Fi Protected Setup (WPS) PIN Attacks**: Some routers offer a WPS feature for easy connection setup, which can be vulnerable to brute-force attacks. Attackers attempt to guess the WPS PIN to gain unauthorized access to the network.

Mitigation Strategies:

- **Use Strong Encryption**: Implementing strong encryption protocols like WPA3 is crucial for protecting wireless communications from eavesdropping and other interception attempts.
- **Secure Wi-Fi Access Points**: Regularly change passwords, disable WPS, and hide the SSID to make it harder for attackers to find and exploit your network.

- **Network Segmentation**: Separate sensitive data and devices on different network segments to reduce the risk of unauthorized access.
- **Monitor for Unauthorized Devices**: Regularly scan your network for unauthorized access points or devices that could indicate a breach.
- **Educate Users**: Raise awareness among network users about the risks of connecting to unknown wireless networks and the importance of using secure, encrypted connections.

By understanding the unique security challenges that wireless networks face and implementing robust security practices, organizations and individuals can significantly reduce their vulnerability to these common wireless attacks.

Auditing and Monitoring: These processes help maintain network security. Auditing involves assessing network configurations, identifying vulnerabilities, and ensuring compliance with security policies. Continuous monitoring provides real-time insights into network traffic and behavior, allowing for the rapid detection of anomalies and potential security incidents. Network auditing tools and practices, combined with vigilant monitoring, form the backbone of effective network security practices.

Network Security Tools and Solutions: The world of network security is equipped with an array of specialized tools and solutions designed to bolster the defense of digital assets and data. These tools play a huge role in identifying vulnerabilities, detecting threats, and responding to security incidents. Network administrators and security specialists should understand their functionalities and applications.

Firewalls examine incoming and outgoing traffic and enforce access policies. Next-generation firewalls (NGFWs) incorporate advanced features like application-layer filtering and deep packet inspection for enhanced security. Antivirus software scans for and removes malicious software (malware) from networked devices, preventing infections from spreading.

Security information and event management (SIEM) solutions provide centralized log management and real-time analysis of security events. They enable security professionals to correlate information from multiple sources to identify potential threats. Additionally, SIEM tools assist in compliance reporting, a critical aspect for organizations subject to regulatory requirements.

Network monitoring solutions offer comprehensive visibility into network performance and security. They track bandwidth utilization, traffic patterns, and device behavior. Monitoring tools help administrators detect anomalies and respond promptly to issues that could compromise network integrity.

Cryptography and Secure Communication

Cryptography is like a secret code for digital information. It's a way to make sure that when you send a message or data over the internet or any network, it stays safe from prying eyes. Think of it as putting your message in a locked box that only the intended recipient can open. Cryptography is the foundation of secure communication online.

Core Cryptographic Concepts: In cryptography, we use special math and techniques to protect information. The two main concepts are encryption and decryption. Encryption turns your message into a secret code that's useless to anyone who intercepts it without the key to unlock it. Decryption is the process of turning that secret code back into the original message. We use keys to lock and unlock the code, and these keys are like super-secret passwords.

Implementing Secure Communication: Secure communication happens all around us. When you visit a website with "https://" in the address, it means your connection is secure. This is thanks to protocols like SSL/TLS. Email encryption keeps your emails private, and secure file transfer methods protect your data when you send it to others. These practices ensure that your online interactions remain safe from eavesdroppers.

Public Key Infrastructure (PKI): PKI is like a system of trust on the internet. It helps ensure that the person or website you're communicating with is really who they claim to be. PKI uses digital certificates issued by trusted organizations called Certificate Authorities (CAs). These certificates act like digital IDs, proving that a website or person is legitimate. When you see a little padlock icon in your web browser, that's PKI in action, assuring you that your connection is secure.

Challenges in Cryptography: Cryptography is continuously evolving because hackers and technology change too. One big challenge is quantum computing, which has the potential to break current encryption methods. Researchers are working on new ways to secure data against quantum threats. Another challenge is staying ahead of cybercriminals who are always finding new ways to attack.

That's why it's important to keep up with the latest security measures to protect digital information.

Advanced Security Considerations

This section delves into advanced cybersecurity practices and strategies that extend beyond the basics. It covers topics such as data protection, user awareness, and security for cloud computing, IoT, and mobile devices.

Data Protection and Privacy

In the digital age, data is everywhere, and it's paramount to keep it safe. Data protection and privacy ensure that your personal and sensitive information is secure. Whether you're sharing data online, using a mobile app, or working with an organization, understanding these concepts helps you make informed choices about your data.

Data Security Fundamentals: Data security is like a lock and key for information. It begins with classifying data based on its importance and sensitivity. Access control ensures that only authorized individuals can access specific data. Encryption transforms data into a secret code, making it unreadable to anyone who doesn't have the key. These fundamentals create layers of protection around your data.

Privacy Laws and Regulations: Governments worldwide have implemented laws to safeguard individuals' data rights. GDPR (General Data Protection Regulation) and CCPA (California Consumer Privacy Act) are examples. They give you control over your data, including the right to know what's collected and the ability to request its deletion. Understanding these laws empowers you to exercise your data privacy rights.

Data Breaches and their Impact: Data breaches occur when unauthorized individuals access and steal sensitive data. They can happen to companies or individuals and can lead to identity theft, financial loss, or reputational damage. Knowing the signs of a data breach and how to respond is important for minimizing the harm.

Protecting Personal Data: Everyone can take steps to protect personal data. This includes using strong, unique passwords for online accounts, enabling two-factor authentication, and being cautious when sharing personal information online. Organizations can also implement data protection measures, such as encryption and regular security updates.

Data Privacy in the Workplace: Balancing data privacy in the workplace is a necessity. Employees have rights regarding their data, such as the right to know if their data is collected and how it's used. Employers must strike a balance between monitoring for security purposes and respecting employee privacy. Clear data privacy policies help achieve this balance.

Security in Cloud Computing

Cloud computing has transformed the way we access and manage data and applications. It offers on-demand access to a shared pool of computing resources over the internet. However, this convenience comes with security considerations. Cloud computing security encompasses protecting data, applications, and infrastructure hosted in the cloud. This involves safeguarding sensitive information, ensuring data privacy, and defending against cyber threats.

Security measures in cloud computing are complex. They include robust authentication mechanisms to verify the identity of users and applications accessing cloud resources. Encryption plays a massive role in securing data, both in transit and at rest, making it unreadable to unauthorized parties. Regular security updates and patch management can help in addressing vulnerabilities and protecting against evolving threats.

Cloud Deployment Models: Cloud deployment models determine how cloud resources are made available to users and organizations. They include public, private, and hybrid clouds. Public clouds are owned and operated by third-party cloud service providers (CSPs) and offer scalable computing resources on a pay-as-you-go basis. They are cost-effective and suitable for businesses seeking flexibility and scalability without the burden of managing infrastructure.

Private clouds, on the other hand, are dedicated to a single organization and may be hosted on-premises or by a third-party provider. They offer enhanced control, security, and customization but require more significant management and resources. Hybrid clouds combine elements of both public and private clouds, enabling data and applications to move seamlessly between environments. This hybrid approach provides flexibility while allowing sensitive data to remain on private infrastructure. The choice of cloud deployment model depends on an organization's specific needs, security requirements, and existing IT infrastructure.

Shared Responsibility Model: The shared responsibility model is a fundamental concept in cloud security. It defines the division of security responsibilities between CSPs and cloud users. CSPs handle securing the underlying cloud infrastructure, including data centers, servers, and network architecture. They implement physical security measures, maintain network availability, and manage data center access controls.

On the other hand, cloud users are accountable for securing their data, applications, and configurations within the cloud environment. This includes defining user access controls, encrypting sensitive data, and implementing security best practices for cloud-based applications. It is recommended to understand this shared responsibility model to ensure that all aspects of cloud security are adequately addressed.

Common Cloud Security Threats: Cloud computing introduces unique security challenges. Common cloud security threats include data breaches, insider threats, and misconfigurations. Data breaches occur when unauthorized individuals gain access to sensitive data stored in the cloud. Insider threats can be intentional or unintentional and may involve employees or contractors who misuse their access privileges.

Misconfigurations, such as improperly configured access controls or firewall rules, can expose cloud resources to potential attacks. Other threats include distributed denial of service (DDoS) attacks, where attackers overwhelm cloud services with traffic to disrupt operations. Staying vigilant and implementing security measures will help to mitigate these threats and protect cloud resources.

Cloud Security Best Practices: Implementing security best practices is mandatory for maintaining a robust security posture in the cloud. Access control mechanisms, such as role-based access control (RBAC) and MFA, help restrict unauthorized access to cloud resources. Encryption safeguards data from unauthorized interception, both during transit and while stored in cloud repositories.

Continuous monitoring and logging provide real-time visibility into cloud environments, allowing organizations to detect and respond to security incidents promptly. Regular security assessments and vulnerability scans help identify and remediate potential weaknesses. Additionally, cloud users should stay informed about emerging threats and security updates provided by CSPs.

Cloud Compliance and Regulations: Cloud computing often involves the processing and storage of sensitive data, making compliance with data protection regulations a requirement. Regulations like the General Data Protection Regulation (GDPR) and the Health Insurance Portability and Accountability Act (HIPAA) impose specific requirements on how personal and healthcare data is handled.

Cloud users must understand their responsibilities in ensuring compliance within the cloud environment. This includes data classification, encryption, access controls, and audit trails to demonstrate compliance. Non-compliance with these regulations can result in severe penalties, making it necessary for organizations to align their cloud practices with relevant regulations.

Cloud Security Tools and Services: Cloud service providers and third-party vendors offer a range of security tools and services to enhance cloud security. Identity and access management (IAM) solutions allow organizations to define and enforce user roles and permissions, ensuring that only authorized individuals can access resources.

Threat detection and response services employ advanced analytics and machine learning to identify unusual patterns or suspicious activities within the cloud environment. These services enable rapid incident response and mitigation. Cloud providers also offer security certifications and compliance assessments to help organizations assess their security posture and align with industry best practices.

Mobile Security Challenges and Strategies

Mobile devices, such as smartphones and tablets, have become an integral part of our daily lives, offering convenience, connectivity, and productivity. However, the widespread adoption of mobile technology has also made these devices lucrative targets for cybercriminals. Mobile security is the practice of protecting these devices and the data they contain from various threats, ranging from malware and phishing attacks to physical theft or loss.

Mobile security is essential because these devices often store sensitive information, including personal photos, messages, financial data, and even access to business applications. As such, securing mobile devices is not just a matter of protecting hardware; it's about safeguarding personal and professional information.

Common Mobile Security Threats: Mobile devices face a diverse range of security threats. Malware, including viruses, Trojans, and spyware, can infect mobile devices through malicious apps or compromised websites. Phishing attacks attempt to trick users into revealing sensitive information by impersonating trusted sources. Insecure Wi-Fi networks, often found in public places, can expose mobile devices to eavesdropping and data interception. Additionally, device theft or loss can result in unauthorized access to personal data. Understanding these common threats is crucial for mobile users, as it enables them to recognize potential risks and take proactive measures to protect their devices and data.

Securing Mobile Devices: Securing mobile devices starts with basic yet important steps. Setting strong passcodes or using biometric authentication, such as fingerprint or facial recognition, adds an additional layer of security. Enabling device encryption ensures that data stored on the device is protected even if it falls into the wrong hands. Additionally, the ability to remotely wipe and track a lost or stolen device provides peace of mind.

Mobile security is not just about defense; it's also about preparation. Regularly updating the device's operating system and applications is essential to patch vulnerabilities and keep security up to date. By implementing these security measures, mobile users can significantly reduce the risk of unauthorized access to their devices and data.

Mobile App Security: Mobile apps play a key role in the functionality and entertainment of mobile devices. However, not all apps are created equal when it comes to security. App permissions, which dictate what resources an app can access, are important to understand. Users should carefully review permissions and consider whether they are necessary for the app's functionality. App stores, such as the Apple App Store and Google Play Store, provide curated platforms for app distribution, but malicious apps can still slip through the cracks. Users should stick to reputable app stores and exercise caution when considering apps from third-party sources.

Mobile Device Management (MDM): MDM solutions are used by organizations looking to manage and secure mobile devices within their networks. MDM enables IT administrators to monitor and control devices, enforce security policies, and remotely manage device configurations. This level of control contributes to protecting sensitive business data and ensuring compliance with security standards.

MDM solutions offer features like remote device lock and wipe, app management, and secure email access. They can also help organizations track device inventory and enforce encryption policies.

For both businesses and individuals, understanding the role of MDM in mobile security will assist in effectively managing and securing mobile devices.

Secure Mobile Browsing: Secure web browsing is fundamental for preventing mobile device vulnerabilities and data exposure. Users should ensure they are connecting to secure and trusted websites, particularly when entering personal or financial information. Utilizing HTTPS (Hypertext Transfer Protocol Secure) connections adds an additional layer of encryption to web interactions, protecting data in transit. Mobile browsers offer security features like pop-up blockers and privacy modes that enhance protection against online threats. Additionally, users should stay vigilant and avoid downloading files or clicking on links from suspicious or unknown sources.

Mobile Payment Security: Mobile payment options, including NFC (Near Field Communication) payments and mobile wallets like Apple Pay and Google Pay, offer convenience but also introduce security considerations. Secure transactions involve tokenization, which replaces card information with unique tokens for each transaction. Users should ensure their mobile payment methods are protected by strong authentication, such as PINs or biometrics. It's essential to use official payment apps and trusted merchants for mobile transactions. Additionally, monitoring transaction histories and setting up transaction alerts can help users quickly identify any unauthorized activity.

IoT Security: Risks and Mitigations

Internet of Things (IoT) devices are everyday objects that are embedded with sensors, software, and connectivity capabilities, enabling them to collect, exchange, and act upon data. These devices are part of a vast network where physical objects, such as household appliances, wearable technology, industrial machines, and even vehicles, are connected to the internet or other networks. The primary purpose of IoT devices is to enhance functionality, efficiency, and automation in various aspects of our lives.

IoT devices can range from smart thermostats and fitness trackers to connected cars and industrial sensors. They play a pivotal role in the modern digital ecosystem, enabling data-driven decision-making and automation. However, their connectivity also introduces security and privacy considerations, as they become potential targets for cyberattacks.

IoT Security: The IoT represents a paradigm shift in the way devices and objects interact with each other and with us. These devices are equipped with various sensors, connectivity options, and computational capabilities, allowing them to collect data, communicate over networks, and perform actions autonomously. They are integrated into our homes, industries, healthcare systems, and transportation, among others, enhancing convenience and efficiency. However, the increasing proliferation of these devices also introduces security challenges. They become attractive targets for cybercriminals seeking to exploit vulnerabilities for various purposes, including data theft, unauthorized access, and even launching large-scale attacks. It is imperative to ensure the security of these interconnected devices to protect privacy, data integrity, and overall system reliability.

Security Challenges: IoT security presents unique challenges due to the sheer scale of deployment and the diverse range of devices involved. One challenge is the lack of standardized security measures across manufacturers and vendors. This diversity results in inconsistencies in security practices and leaves many devices vulnerable to exploitation. Additionally, IoT devices often have limited computational power and memory, making it challenging to implement robust security features.

The vast number of smart devices creates a massive attack surface for cybercriminals to target. Securing these devices can be complex and costly, especially for organizations managing extensive IoT deployments. The evolving nature of these ecosystems further complicates security efforts, as new vulnerabilities and threats continuously emerge. Addressing these challenges requires a comprehensive approach that involves manufacturers, consumers, and regulatory bodies.

Common Security Threats: IoT devices face a range of security threats, including data breaches, unauthorized access, and IoT-specific attacks. Unauthorized access occurs when malicious actors gain control of a smart device, potentially using it as a gateway to infiltrate a network or compromise other devices. Data breaches involve the exposure or theft of sensitive information transmitted or stored by these devices. IoT-specific attacks include Distributed Denial of Service (DDoS) attacks orchestrated by botnets composed of compromised smart devices. These attacks can disrupt services and networks on a large scale. Other threats include man-in-the-middle attacks, where attackers intercept and manipulate data exchanged between IoT devices and the central system. Understanding these threats is important for implementing effective security measures.

Securing IoT Devices: Securing these devices begins with robust device authentication. Strong authentication mechanisms, such as unique device identifiers and secure communication protocols, ensure that only authorized devices can connect to the network. Firmware updates play a crucial role in addressing vulnerabilities and should be regularly applied to keep devices secure. Network segmentation, isolating IoT devices from critical systems, helps contain potential breaches. Implementing firewalls and intrusion detection systems adds an extra layer of defense. Additionally, device manufacturers should follow security-by-design principles, incorporating security measures during the product development phase. Regular security audits and penetration testing can help identify and mitigate vulnerabilities.

IoT Security Best Practices: Both consumers and organizations can follow best practices to enhance IoT security. Strong password management is instrumental; users should change default passwords, use unique and complex passwords, and enable multi-factor authentication where possible. Regularly updating device firmware is needed to patch known vulnerabilities. Monitoring device behavior for unusual activities can help detect security incidents early. Privacy should be a consideration, and users should review privacy settings and permissions associated with IoT devices. Additionally, individuals and organizations should stay informed about security updates and vulnerabilities related to their devices.

Privacy Concerns in IoT: Smart devices often collect a vast amount of data, including personal and sensitive information. Privacy concerns arise regarding how this data is collected, stored, and used. Users must be aware of data collection practices and provide informed consent. Compliance with data protection regulations, such as the General Data Protection Regulation (GDPR), is required when handling personal data in IoT ecosystems.

Data encryption and anonymization can help protect privacy by ensuring that collected data is secure and not directly linked to individuals. Transparency in data practices, including clear privacy policies and user controls, can address many privacy concerns. Organizations must navigate the balance between collecting data for valuable insights and respecting individuals' privacy rights.

User Education and Security Awareness

In cybersecurity, the human element is often the most vulnerable. Educating users about security practices is not just about imparting knowledge; it's about cultivating a culture of security. It's necessary for individuals to understand the risks associated with their online activities and the steps they can take to mitigate these risks. For instance, users should be trained to identify phishing attempts, understand the importance of regular software updates, and the repercussions of weak passwords. Regular training sessions, updated security guidelines, and interactive learning modules can significantly enhance a user's ability to prevent, identify, and respond to security threats.

Real World Example: In 2017, a large-scale ransomware attack known as WannaCry affected organizations worldwide, including the UK's National Health Service (NHS). The attack exploited vulnerabilities in outdated Windows systems, encrypting data and demanding ransom payments. A significant factor contributing to the attack's success was the lack of regular software updates and user awareness regarding cybersecurity. The NHS had numerous unpatched systems and staff were not adequately trained to recognize and avoid malicious emails that led to the ransomware's spread. The WannaCry incident highlights the critical importance of regular system updates and comprehensive user education in preventing similar cyber-attacks. For more details on this case, see the citation in the Appendix.

Common Threats and Social Engineering: Cyber threats have evolved substantially, and one of the most insidious forms is social engineering. Social engineering exploits human psychology, rather than technical hacking techniques, to gain access to systems, networks, or physical locations. Examples include phishing, where attackers masquerade as a trustworthy entity in an email, or baiting, where they use a false promise to pique a victim's greed or curiosity. Understanding these threats requires knowledge of how attackers operate and the common signs of a social engineering attack, such as unsolicited requests for confidential information or pressure to bypass standard procedures.

Recognizing Suspicious Activity: The ability to recognize suspicious activity is a huge component of cybersecurity. Suspicious activities can range from unexpected pop-ups on a computer to an unknown device connecting to a network. Key indicators include system performance issues, unauthorized software installations, and unexpected requests for authentication. Users should be trained to report these signs immediately to IT professionals. This proactive stance helps in early detection of potential security breaches, minimizing damage and preventing large-scale attacks.

Password Security and Authentication: Password security is a fundamental aspect of cybersecurity. Strong, unique passwords are the first line of defense against unauthorized access. Users should be educated on creating complex passwords that use a mix of letters, numbers, and symbols, and the dangers of using easily guessable passwords like '123456' or 'password'. Furthermore, the importance of regular password changes and the use of MFA where possible enhances security. MFA adds an additional layer of security by requiring two or more verification methods, which significantly reduces the risk of compromise.

Case Study: In 2012, LinkedIn experienced a major security breach where approximately 6.5 million user passwords were stolen and posted on a Russian hacker forum. The breach occurred due to LinkedIn's use of unsalted SHA-1 hash for password encryption, a method vulnerable to attacks. Many users had weak, easily guessable passwords, which compounded the problem. The incident underscored the necessity of strong, unique passwords and advanced encryption techniques like salting and hashing. Following the breach, LinkedIn enhanced its security measures, including implementing stronger password policies and two-factor authentication to protect user accounts against similar attacks. For more details on this case, see the citation in the Appendix.

Security Policies: Organizational security policies are the backbone of a company's cybersecurity defense. They provide a framework for handling and protecting sensitive information and outline the responsibilities of employees. Users should be familiar with these policies, which cover areas like acceptable use, data handling, and incident response. Regular training and updates on these policies ensure that all employees understand their role in safeguarding the organization's digital assets and about the importance of understanding and adhering to these policies to maintain the overall security of the organization.

Cybersecurity Resources and Training: IT pros should stay updated with the latest cybersecurity trends and knowledge. Users should be encouraged to participate in ongoing cybersecurity training and awareness programs. These resources can include online courses, workshops, webinars, and industry conferences. Also, staying informed about the latest cybersecurity news, understanding emerging threats, and learning about new security technologies and practices can greatly enhance an individual's ability to respond effectively to cyber threats. Consult the checklist below for fundamental recommendations to secure your devices.

Cybersecurity Awareness Checklist:

1. **Recognize Phishing Emails:**
    - Check the sender's email address for authenticity.
    - Look for generic greetings and spelling errors.
    - Be wary of urgent or alarming content that requests immediate action.
2. **Secure Password Practices:**
    - Use complex passwords with a mix of letters, numbers, and symbols.
    - Avoid using the same password across multiple accounts.
    - Consider using a reputable password manager. i.e. LastPass, Bit Warden
3. **Update Software Regularly:**
    - Keep all devices and software up to date with the latest security patches.
    - Enable automatic updates where possible.
4. **Safe Internet Browsing:**
    - Use secure, reputable websites, especially when entering personal information. Utilize HTTPS instead of HTTP
    - Avoid clicking on unknown links or downloading files from untrusted sources.
5. **Personal Device Security:**
    - Install antivirus and anti-malware software.
    - Use secure Wi-Fi networks and avoid public Wi-Fi for sensitive transactions. Utilize a VPN if necessary.
6. **Data Privacy Awareness:**
    - Be cautious about the personal information shared on social media.
    - Understand privacy settings on various online platforms.
7. **Reporting Suspicious Activities:**
    - Know how to report potential security threats in your workplace or personal devices.
    - Stay informed about the latest cybersecurity threats and trends.

Cybersecurity is not a static field; it's a dynamic and ever-evolving challenge. This chapter underscores the vital importance of staying vigilant and educated in the face of growing cyber threats. From understanding the nuances of social engineering to implementing robust password policies, and securing personal devices to adhering to organizational security policies, each aspect contributes to a comprehensive defense strategy. The power of knowledge in cybersecurity cannot be overstated – it empowers individuals and organizations to not only react to threats but also to anticipate and prevent them. As technology continues to advance, so must our commitment to cybersecurity education and awareness, ensuring a safer digital environment for all. In the next chapter, we will dig into more advanced cybersecurity topics.

Reflect and Apply Questions

1. **Recognizing Online Scams:** Think of a time when you received a suspicious email or message. What were the red flags, and how did you respond? Relate this to the importance of recognizing cybersecurity threats.
2. **Tech Support in Everyday Life:** Think of a time when a friend or family member asked for your help with a tech issue (like setting up a device or fixing an internet connection). How did you approach solving this problem, and what does it teach you about troubleshooting in IT?
3. **Embracing Digital Change:** Reflect on your experience when a commonly used app or service updated significantly. How did you adapt to the new interface or features, and what does this say about embracing change in the IT world?
4. **Online Privacy Measures:** Consider your social media usage. How do you ensure your personal information is secure and private? Relate this to the importance of privacy and security in the tech industry.
5. **Learning from Tech Mistakes:** Recall a time when you made an error with a technology or device (like deleting important files or falling for a phishing email). What did you learn from this mistake, and how can this lesson be applied in an IT career?
6. **Adapting to New Technologies:** Think about a time when you had to learn a new technology or software for personal use. How did you approach learning it, and how does this experience relate to staying updated with IT trends?

Chapter 9: Cybersecurity: Beyond the Basics

In Chapter 9 we discuss some of the more advanced aspects of cybersecurity, indispensable for anyone looking to deepen their understanding of this field. This chapter aims to equip readers with knowledge about sophisticated strategies, tools, and practices that go beyond foundational concepts. We explore the complexities of Identity and Access Management (IAM), the intricacies of combating advanced malware and viruses, and the importance of adhering to security compliance and standards. Additionally, this chapter sheds light on ethical hacking, penetration testing, and the vital roles they play in strengthening cyber defenses. We will also address how to effectively respond to incidents and recover from disasters, the impact of AI and machine learning in cybersecurity, and the emerging technologies that are shaping the future of cyber threats. Understanding these advanced topics assists in building a robust, security-focused culture within organizations and for navigating a career in the world of cybersecurity.

Advanced Security Practices

This section dives into advanced techniques and technologies used in cybersecurity to proactively defend against evolving threats. It explores ethical hacking and penetration testing methodologies, providing insights into how organizations can identify vulnerabilities in their systems and networks before malicious actors exploit them. Additionally, it discusses the role of artificial intelligence and machine learning in enhancing cybersecurity defenses, including their applications in threat detection, anomaly detection, and automated response.

Ethical Hacking and Penetration Testing

Ethical hacking, often glamorized in movies and TV shows, is a legitimate and legal cybersecurity practice. Unlike the stereotypical portrayal of hackers as cybercriminals, ethical hackers use their skills for constructive purposes. They are security experts authorized to penetrate networks and computer systems with the aim of finding and fixing security vulnerabilities. This process involves the same techniques used by malicious hackers, but the intent is to improve and secure systems rather than to exploit them. Ethical hacking is about understanding the mindset and tactics of attackers to better defend against them.

In the real world, ethical hacking is a structured, methodical process, not the fast-paced hacking often depicted in popular media. Ethical hackers follow strict ethical guidelines and legal frameworks, ensuring their actions benefit the organization without causing harm or breaching trust. Their work is critical in identifying and addressing security weaknesses before they can be exploited by malicious actors.

The Role of a Penetration Tester: A penetration tester, or pen tester, simulates cyber-attacks to identify vulnerabilities in computer systems, networks, and applications. They use a variety of tools and techniques to probe for weaknesses, such as unpatched software, insecure configurations, and weak passwords. The primary goal is to discover these vulnerabilities before malicious hackers do, providing valuable insights into potential security gaps. Penetration testers must have a deep understanding of both technology and cybersecurity. They often think outside the box, adopting an attacker's mindset to uncover hidden weaknesses. Their findings are then documented in detailed reports, which are used by organizations to strengthen their cybersecurity defenses.

Common Tools and Techniques: Ethical hackers and penetration testers use a range of tools and techniques to assess the security of systems. Vulnerability scanners automate the process of identifying security weaknesses in networks and applications. Network sniffers allow testers to capture and analyze network traffic, identifying potential data leaks and protocol vulnerabilities. Penetration testing software, like Metasploit, provides frameworks for developing and executing exploit code against identified vulnerabilities.

The Legal and Ethical Framework: Ethical hacking is governed by a strict legal and ethical framework. Essential to this practice is obtaining explicit permission from the organization that owns the system. Ethical hackers must adhere to laws and regulations, ensuring their activities remain within legal boundaries. They also follow a code of ethics, which includes respecting privacy, not exploiting vulnerabilities for personal gain, and maintaining confidentiality.

Steps in a Penetration Testing Process:

Here are some common steps in a penetration testing process, each vital for a thorough security assessment:

1. **Planning and Reconnaissance:** This initial stage involves defining the scope and goals of a penetration test. Testers gather information about the target system, such as network and domain names, to understand how it operates and its potential vulnerabilities.
2. **Scanning:** Involves using tools to scan the target system for known vulnerabilities. This might include network scanning to identify live hosts and open ports, and vulnerability scanning to detect security weaknesses.
3. **Gaining Access:** This step involves attempting to exploit identified vulnerabilities, typically using techniques like SQL injection, cross-site scripting, or buffer overflows, to gain unauthorized access to the system.
4. **Maintaining Access:** The goal here is to see if the vulnerability can be used to achieve a persistent presence in the exploited system, simulating advanced persistent threats.
5. **Analysis and Reporting:** The final step involves analyzing the data gathered during the test and preparing a detailed report. This report outlines the vulnerabilities discovered, the extent to which the tester was able to exploit them, and the potential impact. It also includes recommendations for mitigating the identified risks, helping the organization to strengthen its defenses against actual cyberattacks.

Building a Career in Ethical Hacking: Building a career in ethical hacking requires a blend of education, skills, and certifications. Aspiring ethical hackers should start with a strong foundation in IT, especially in areas like networking and system administration. Gaining certifications such as Certified Ethical Hacker (CEH) or Offensive Security Certified Professional (OSCP) is highly beneficial. These certifications not only provide valuable knowledge and skills but also demonstrate credibility in the field.

Practical experience is paramount in ethical hacking. Beginners can start with practice labs, hackathons, and participating in bug bounty programs to hone their skills. Bug bounty programs are offered from many companies big and small where they will pay ethical hackers if they can find bugs in their systems. Of course, there are rules that need to be followed and some items considered off limits. Networking with other cybersecurity professionals and staying updated with the latest trends and tools are also key steps in building a successful career in ethical hacking.

Real-World Example: An example of ethical hacking in action can be seen in a penetration test conducted for a financial institution. The testers identified vulnerabilities in the web application that could allow unauthorized access to sensitive customer data. By exploiting these weaknesses, they demonstrated the potential impact of a real cyber-attack. The institution used the findings to patch vulnerabilities and improve its security posture, showcasing the practical benefits of ethical hacking.

Incident Response and Disaster Recovery

An incident, in the context of cybersecurity, can be broadly defined as any event that threatens the integrity, confidentiality, or availability of information assets. This can range from external attacks like hacking or phishing to internal threats such as data leaks or system failures. Disasters include larger-scale events that significantly disrupt operations, like natural disasters impacting data centers, or widespread malware infections. IT experts should understand the varied nature of incidents and disasters. An effective response plan recognizes these potential threats and outlines specific actions to address them, minimizing their impact on the organization.

Phases of Incident Response:

1. **Identification:** This involves detecting and determining the nature of the incident. It includes monitoring systems for signs of a security breach, such as unusual network traffic or unauthorized access attempts. Prompt and accurate identification is key to a successful response.
2. **Containment:** Once an incident is identified, the immediate goal is to contain it. This might involve isolating affected systems to prevent the spread of the threat and securing network segments.
3. **Eradication:** After containing the incident, the next step is to remove the threat. This includes eliminating malware, closing security gaps, and addressing vulnerabilities that led to the incident.
4. **Recovery:** In the recovery phase, affected systems and services are restored to their normal operations. This involves careful planning to avoid re-introducing the threat and may include data restoration from backups.

5. **Lessons Learned:** This final phase involves analyzing the incident and the response to it. It's a critical step for improving future response efforts, updating policies, and training based on the insights gained.

Incident Response Teams: Incident response teams play a major role in managing cybersecurity incidents. Beyond identifying and analyzing security breaches, these teams coordinate response efforts across various departments. They also work on implementing recovery measures. Effective incident response requires a blend of technical skills for understanding and mitigating threats, strong problem-solving abilities for quickly devising solutions, and excellent communication skills for disseminating information within the organization and to external stakeholders, as needed.

Testing and Improving Response Plans: It's not enough to have a response plan in place; it must be actively tested and updated. Regularly scheduled drills and simulations of various types of cyber incidents are mandatory for evaluating the effectiveness of these plans. After each test or real incident, a thorough review should be conducted to identify any weaknesses or areas for improvement in the response plan. This process of continuous improvement ensures that the organization remains prepared for new and evolving cyber threats.

Legal and Regulatory Considerations: Navigating the legal and regulatory aspects of incident response is complex but important. Organizations must be aware of the legal requirements related to reporting data breaches, which can vary depending on the region and the nature of the data involved. Compliance with industry-specific regulations, such as those in finance or healthcare, is also required. Failure to comply can result in substantial legal penalties, not to mention damage to reputation and customer trust.

Technology and Tools: In incident response, technology and tools are used for effectively managing and resolving security incidents. Advanced forensic tools are used to investigate and analyze breaches, uncovering how the attack occurred and the extent of the damage. Communication systems are used for coordinating response efforts, especially in large organizations where rapid dissemination of information is required. Incident tracking solutions provide a structured approach to documenting incidents, tracking their resolution and ensuring that lessons are learned for future improvements. These tools assist in the immediate response and in the longer-term analysis and strengthening of cybersecurity measures. Modern organizations should ensure the integration of such technologies into incident response strategies to combat sophisticated cyber threats.

AI and Machine Learning in Cybersecurity

Artificial Intelligence (AI) and Machine Learning (ML) are revolutionizing cybersecurity, offering new ways to combat increasingly sophisticated cyber threats. AI in cybersecurity involves using intelligent algorithms to detect patterns, analyze risks, and make decisions, often in real time. Machine Learning, a subset of AI, focuses on the ability of systems to learn and improve from experience without being explicitly programmed. These technologies are being employed to automate complex processes for detecting and responding to cyber threats, greatly enhancing the efficiency and effectiveness of cybersecurity measures.

Applications of AI in Cybersecurity: AI is used for anomaly detection, where AI algorithms analyze network traffic to identify unusual patterns that may indicate a breach. In threat intelligence, AI helps in processing vast amounts of data to identify emerging threats. In automated incident response, AI enables rapid response to threats by automating certain decision-making processes. AI-driven security systems are becoming essential in proactively identifying and neutralizing cyber threats.

Machine Learning Techniques in Threat Detection: By analyzing historical data, ML algorithms can learn to identify patterns and anomalies that signify potential security incidents. For instance, ML can be used to detect subtle signs of ransomware or to identify new forms of malware based on deviations from normal behavior. These capabilities allow for more proactive and predictive cybersecurity measures, moving beyond traditional reactive approaches.

Challenges and Limitations: While AI and ML offer significant advantages in cyber, they also come with challenges. The accuracy of these systems heavily depends on the quality and quantity of the data used for training. There's also a need for continuous human oversight to interpret AI decisions and intervene when necessary. Additionally, AI systems themselves can be targets of cyber-attacks, where attackers manipulate the data or algorithms to evade detection or cause false positives.

Ethical Considerations: The use of AI in cybersecurity raises important ethical considerations. Privacy concerns emerge when AI systems process large amounts of personal data. There's also the risk of bias in AI algorithms, which can lead to unfair or discriminatory outcomes. Furthermore, the ethical implications of using AI for offensive cybersecurity measures, such as automated hacking back, need careful consideration to avoid unintended consequences.

Future of AI in Cyber: The field is moving towards more autonomous security systems capable of making complex decisions and responding to threats in real-time. The integration of AI is expected to become deeper, with more advanced algorithms capable of handling emerging threats like deepfakes or AI-generated attacks. However, the cyber community must also prepare for adversarial AI, where attackers use AI to develop more sophisticated attack strategies.

Organizational Security Strategies

In this section, we will focus on organizational level cyber-security measures. We'll examine (IAM) principles and practices for ensuring secure access to resources and data within an organization. Then we dive into the importance of fostering a security-focused culture within organizations, emphasizing the role of leadership, policies, training, and employee awareness in promoting cyber best practices and mitigating risks effectively.

Identity and Access Management (IAM)

IAM is a framework for business processes that facilitates the management of electronic identities. The core objective is to ensure that the right individuals access the right resources at the right times and for the right reasons. This involves verifying each user's identity, determining the appropriate level of access, and managing user access to networks and data. IAM systems are a critical component in safeguarding an organization's digital assets against unauthorized access, thereby playing a major role in the overall cybersecurity strategy.

Types of IAM Systems: IAM systems vary in complexity and scope, ranging from on-premises solutions to cloud-based services. On-premises IAM solutions are hosted within the organization's own data center, offering a high level of control but requiring significant maintenance and infrastructure. Cloud-based IAM, on the other hand, offers scalability and accessibility, with the service provider managing the infrastructure. A hybrid approach combines both, offering flexibility. Main features of IAM systems include

Single Sign-On (SSO), where users access multiple applications with one set of credentials, and Multi-Factor Authentication (MFA), which adds additional layers of security.

IAM in Cybersecurity: By managing user identities and controlling access to resources, IAM systems can effectively minimize the risk of data breaches and cyber-attacks. They ensure that only authorized users gain access to sensitive systems and data, and track user activities, which is necessary for detecting potential security incidents. In a landscape where threats are increasingly sophisticated, having a robust IAM is indispensable for protecting an organization's information assets.

IAM Best Practices: Effective IAM involves several best practices. Regularly reviewing and updating access privileges ensures that users have the appropriate level of access for their current role. Implementing the principle of least privilege minimizes each user's exposure to sensitive data and systems. Educating users on safe password practices is also crucial, as compromised credentials are a common cause of security breaches. Additionally, regular audits and compliance checks help in maintaining a strong IAM system.

Challenges and Future Trends in IAM: Managing IAM can be challenging, particularly in complex environments with diverse user groups and regulatory requirements. Balancing security with user convenience is often a difficult issue. Looking forward, the field of IAM is evolving rapidly, with biometric authentication and artificial intelligence being integrated into more systems. These advanced technologies promise enhanced security, user experience, and efficiency. However, they also bring challenges, such as privacy concerns and the need for more sophisticated security protocols. Modern organizations should stay ahead of these trends to protect their assets effectively in a constantly changing digital landscape.

Case Study: A practical example of IAM's effectiveness can be seen in the case of a financial institution that experienced an attempted data breach. The organization had implemented a robust IAM system with strict access controls and MFA. When attackers attempted to access sensitive customer data using stolen credentials, the MFA requirement thwarted their efforts. This incident demonstrates how IAM, particularly advanced features like MFA, can be a valuable defense against cyber-attacks, safe-guarding both the organization and its customers.

Malware, Viruses, and Threat Mitigation

Malware, short for malicious software, comes in various forms, each with unique characteristics and risks. Viruses, one of the oldest types, attach themselves to clean files and spread throughout a computer system, corrupting files and affecting performance. Worms operate similarly but can spread across networks without human interaction. Trojans disguise themselves as legitimate software, tricking users into installing them. Ransomware is malicious software that encrypts a user's data and demands payment for its release. Lastly, spyware covertly collects user information without their knowledge. Understanding these types of malware helps in implementing specific defensive measures.

Evolution of Cyber Threats: Cyber threats have evolved significantly, becoming more sophisticated over time. Early viruses like the Morris Worm of 1988 caused widespread damage due to its self-replicating nature. The ILOVEYOU virus in 2000 demonstrated the dangers of socially engineered malware. Recent examples include WannaCry, a ransomware attack that affected hundreds of thousands of computers worldwide in 2017. These incidents highlight the evolving nature of cyber threats and underscore the necessity for continuous advancement in cybersecurity measures. For more details on these cases, see the citations in the Appendix.

Detection and Prevention Techniques: Detecting and preventing malware involves a combination of software tools and cybersecurity practices. Antivirus software is used to identify and remove known malware signatures. Firewalls help in monitoring and controlling incoming and outgoing network traffic based on predetermined security rules. Intrusion detection systems (IDS) are used to detect unusual network or system activities that might indicate a breach. Behavioral analysis tools go a step further by analyzing the behavior of software to identify unknown threats. Regular system scans and real-time monitoring are core components of an effective defense strategy.

Best Practices in Threat Mitigation: To mitigate the risk of malware and viruses, organizations and individuals should adhere to several best practices. Keeping software and operating systems up to date is important, as updates often include patches for security vulnerabilities. Practicing safe browsing habits and being cautious about downloads and email attachments can prevent many infections. Regular backups of important data ensure recovery in case of an attack. Users should take email security measures, such as spam filters and caution with unknown senders. Implementing these practices creates multiple layers of defense, significantly reducing the likelihood of successful malware attacks.

Role of User Education: Many cyber-attacks start with a user mistake, such as clicking on a malicious link or downloading a compromised file. Regular training sessions on recognizing phishing emails, safe internet practices, and the importance of strong passwords can significantly reduce the risk of malware infections. An informed user base acts as a first line of defense against cyber threats.

Emerging Threats and Future Preparedness: As technology advances, so do malware techniques. Emerging threats include AI-powered malware capable of learning and adapting to bypass traditional security measures. Future preparedness involves staying informed about new threats, investing in advanced detection technologies, and adopting a proactive approach to cybersecurity. Organizations should conduct regular security audits, update their incident response plans, and engage in continuous staff training to stay ahead of these evolving threats.

Case Study: A major retail chain experienced a sophisticated malware attack targeting their point-of-sale systems. They employed advanced malware detection tools using AI and machine learning, which identified and neutralized the threat. Post-incident, they enhanced their systems with continuous monitoring and regular cybersecurity training for their staff. This case study demonstrates the role of advanced malware protection and employee training in cybersecurity.

Security Compliance and Standards

Security compliance and standards are important components of cybersecurity. They provide a set of guidelines and best practices for protecting information and managing risks. Compliance ensures that organizations adhere to legal and regulatory requirements, while standards offer a framework for implementing effective security measures. Understanding and following these guidelines is mandatory for maintaining the integrity, confidentiality, and availability of data in various industries.

Key Security Standards and Frameworks:

- **ISO 27001:** An international standard for managing information security. It provides a set of policies, procedures, and controls for establishing an Information Security Management System (ISMS).

- **NIST Framework:** Developed by the National Institute of Standards and Technology, it offers guidelines to help organizations manage and reduce cybersecurity risk, focusing on five core functions: Identify, Protect, Detect, Respond, and Recover.
- **General Data Protection Regulation (GDPR):** A legal framework set by the European Union that sets guidelines for the collection and processing of personal information of individuals within the EU.
- **Health Insurance Portability and Accountability Act (HIPAA):** U.S. legislation that provides data privacy and security provisions for safeguarding medical information.

Implementing Compliance: This involves several steps, starting with understanding the applicable standards and regulations. Organizations must conduct thorough risk assessments to identify potential vulnerabilities and develop comprehensive security policies and procedures. To ensure employees understand and adhere to these policies, implement regular training sessions. Ongoing compliance is maintained through periodic audits and by staying updated with changes in standards and regulations.

Challenges in Security Compliance: Organizations often face challenges in achieving and maintaining compliance, such as the complexity of regulations, especially when operating across multiple jurisdictions. The cost of implementation can be weighty, particularly for small to medium-sized enterprises. Additionally, ensuring that all employees are consistently compliant with security policies requires ongoing education and vigilance.

Compliance Audits and Reporting: Compliance audits are critical for assessing an organization's adherence to security standards and regulations. These audits involve evaluating security policies, procedures, and controls to ensure they meet compliance requirements. Effective preparation for audits includes regular internal reviews and maintaining accurate records of compliance efforts. Reporting compliance to regulatory bodies is also a key aspect, often requiring detailed documentation of security practices and incidents.

Building a Security-Focused Culture

A strong security culture is important in any organization, acting as the foundation of comprehensive cybersecurity. It's about instilling a mindset where every employee understands the importance of security and their role in maintaining it. This culture helps in ensuring that security policies are not just formalities but integral parts of everyday work life. A security-focused culture significantly reduces risks of breaches, as employees are more vigilant, follow best practices, and are more likely to report suspicious activities. It's not just about having the right tools and policies; it's about people actively using and supporting them.

Key Elements of a Security Culture: Developing a strong security culture involves several main elements. Leadership commitment is huge; leaders must champion cybersecurity initiatives and model good security practices. Employee awareness and training are also necessary, ensuring that all staff understand the risks and their role in mitigating them. Effective communication is needed to keep security front and center in everyone's mind. Regular updates about potential threats and security protocols help in reinforcing this culture. Lastly, a sense of collective responsibility should be fostered, where security is seen as everyone's business, not just the IT departments.

Strategies for Building a Security Culture: Cultivating a security culture requires proactive strategies. Incorporate regular, engaging training sessions that go beyond mere compliance. These should be interactive and relevant to employees' roles. Creating a security-aware environment through newsletters, posters, and workshops can also be effective. Involving employees in security discussions and decision-making processes can foster a sense of ownership. Recognizing and rewarding good security practices can also motivate employees to take security seriously. All of these strategies contribute to a workplace where security is part of the daily conversation.

Measuring and Monitoring Security Culture: Regular surveys can gauge employee attitudes and knowledge about cybersecurity. Monitoring incident reports and how they are handled provides insights into the practical application of security policies. Feedback mechanisms allow employees to voice concerns and suggest improvements. Tracking these metrics over time helps in understanding the impact of security initiatives and where adjustments may be needed.

Overcoming Challenges: Building a security-focused culture can face challenges like resistance to change or complacency. Overcoming these requires clear communication from leadership about the importance of cybersecurity. Engaging training that resonates with employees' daily activities can help in changing attitudes. Leadership involvement is required; when leaders prioritize and visibly support cybersecurity efforts, it sets a tone for the entire organization. Celebrating successes in improving security also helps in building momentum and buy-in from all staff members.

Integrating Security Culture into Business Processes: Integrating security culture means embedding security considerations into every aspect of the organization's operations. This involves including security checkpoints in project lifecycles, ensuring that new initiatives are evaluated for security risks. Regular dialogues between security teams and other departments can foster a more integrated approach. Security should be a standing item in meetings and decision-making processes, ensuring it is always considered. By making security a fundamental part of business processes, organizations can ensure it is not overlooked.

Emerging Trends and Career Paths

This section explores emerging trends and future directions in cyber. Additionally, it provides brief insights into the dynamic scene of cybersecurity careers, highlighting various career paths, roles, and opportunities available to professionals in the field.

Cybersecurity Trends

Rise of AI and Machine Learning: AI and machine learning are increasingly included in cybersecurity, offering sophisticated approaches to threat detection and response. These technologies enable the analysis of vast datasets for unusual patterns indicative of cyber threats, far beyond the capacity of human analysis. AI-driven security systems are continually learning from new data, improving their ability to identify and respond to threats over time. This evolution represents a significant shift in cybersecurity strategies, as AI and ML can preemptively identify vulnerabilities, predict potential attacks, and automate responses to security incidents. The integration of AI in cyber is not just a trend; it's becoming a necessity in combating the growing sophistication of cyber threats.

Increased Focus on Cloud Security: The migration to cloud-based services has necessitated a stronger emphasis on security. As businesses store more sensitive data in the cloud, the need to protect this data from breaches and unauthorized access has become paramount. This has led to the development of advanced cloud security measures, including enhanced encryption techniques, robust access controls, and continuous monitoring for potential threats. The challenge lies in securing a dynamically changing environment while maintaining user accessibility and compliance with various regulatory standards. Cloud security is evolving into a complex field that requires specialized knowledge and skills, reflecting the importance of protecting data in the cloud era.

IoT and Endpoint Security: The exponential growth of smart devices has introduced new challenges in cybersecurity. These devices often lack the security features of more traditional computing equipment, making them vulnerable to attacks. The issue is compounded by the sheer number of IoT devices and their integration into critical systems, from smart homes to industrial controls. Securing these endpoints requires innovative approaches that consider the unique nature of these devices. Strategies include developing secure IoT frameworks, regular firmware updates, and adopting zero-trust security models to ensure that every device is authenticated and its data transmissions are secure. The security of these devices is indispensable, not only for safeguarding the data they manage but also for ensuring the safety and reliability of the systems they oversee.

Emerging Threats and Attack Vectors: Cybersecurity is a constantly moving target, with new threats emerging regularly. Ransomware attacks have evolved to become more targeted, while phishing scams are now more sophisticated, often leveraging AI to mimic legitimate communications. There's also a rise in attacks on critical infrastructure and supply chain vulnerabilities. IT professionals should understand these emerging threats to ensure the development of effective defense strategies. This requires not only technological solutions but also a comprehensive approach involving employee education and robust security policies.

Remote Workforce Security: The shift to remote work has brought unique cybersecurity challenges. Organizations must secure data across diverse, often uncontrolled environments while ensuring connectivity and employee productivity. This shift necessitates strong remote access solutions, endpoint security, and policies that address the vulnerabilities of remote work setups. Employees should be trained on security best practices for remote work, as the human element often remains the weakest link in cybersecurity.

Skill Gap and Education: There's a growing gap between the demand for skilled cybersecurity specialists and the available talent pool. Addressing this gap is important for businesses and the security of digital infrastructure as a whole. This situation presents a significant opportunity for those willing to acquire the skills needed in this field. Cybersecurity education is evolving, with more courses, certifications, and training programs available than ever before, aimed at equipping the next generation of cybersecurity experts.

Future Predictions and Preparedness: The future of cyber is expected to involve an increased reliance on AI for threat detection and response, further development of quantum-resistant cryptography, and a greater focus on securing smart devices. Organizations must stay ahead of these trends and be prepared for the evolving cybersecurity landscape. This includes adopting the latest technologies and fostering a culture of security awareness and resilience. Staying informed and agile is key in an environment where threats are continuously evolving.

Career Paths in Cybersecurity

The cybersecurity career field is vast and diverse, reflecting the wide range of threats and challenges in the digital world. It offers numerous career paths, from technical roles like network security and threat analysis to strategic positions such as policy development and cybersecurity law. The demand for cybersecurity experts is growing exponentially, driven by the increasing frequency and sophistication of cyber threats. This growth is creating opportunities across various sectors, including government, healthcare, finance, and technology. A career in cyber promises a wide array of roles and offers the potential for rapid career advancement and specialization.

Entry-Level Roles and Progression: Entry-level roles such as Security Analyst, Junior Penetration Tester, or IT Security Administrator, provide a foundational experience in the field. These positions typically involve monitoring networks for security breaches, analyzing threats, and implementing basic protective measures. Progression from these roles can be swift, with opportunities to specialize in areas like forensics, incident response, or ethical hacking. Career advancement often depends on a combination of practical experience, ongoing education, and certification. Many professionals in senior positions have started their careers in these entry-level roles, demonstrating the potential for growth in this field.

Specialized Roles: Specialized roles in cybersecurity cater to specific aspects of the field and require distinct skill sets. Incident Responders are tasked with managing the aftermath of security breaches, requiring quick thinking and problem-solving skills. Cryptographers deal with securing data through encryption, demanding a strong background in mathematics and algorithms. Chief Information Security Officers (CISOs) oversee the entire security posture of an organization, requiring not only technical expertise but also strategic planning and leadership skills. Each specialized role contributes to the overall security infrastructure of an organization.

Skills and Qualifications: A successful career in cyber-security requires a mix of technical, analytical, and soft skills. Technical skills might include network security, ethical hacking, and knowledge of various security tools and software. Certifications such as CompTIA Security+, Certified Ethical Hacker (CEH), or Certified Information Systems Security Professional (CISSP) are highly regarded. Soft skills such as problem-solving, communication, and adaptability are equally important, as cybersecurity experts often need to convey complex information clearly and work under pressure. Continuous learning and staying abreast of the latest trends and technologies are also beneficial for success in this rapidly evolving field.

Education and Training Paths: These paths in cyber-security are varied, allowing individuals to choose a route that best suits their learning style and career goals. Traditional university degrees in cyber, computer science, or related fields provide a solid theoretical foundation. Alternatively, bootcamps and online courses offer more focused, practical training in specific skills. Industry certifications, which can be pursued independently or through professional organizations, are particularly valuable for demon-strating expertise and commitment to the field. The choice of education and training path should align with the individual's career aspirations and the specific skills required in their desired role.

Networking and Professional Development: Attending industry conferences, participating in webinars, and joining professional cybersecurity organizations can provide valuable learning opportunities and connections. Building a professional network can lead to mentorship opportunities, job openings, and collaborations. Actively engaging in the cyber community also keeps professionals informed about the latest trends, challenges, and solutions in the field.

As we conclude Chapter 9, it's clear that the realm of cyber-security is both vast and dynamic. This chapter has explored various advanced topics, from the intricacies of incident response and the complexities of AI in cyber, to the evolving career paths in this field. The knowledge shared here underscores the importance of staying vigilant and informed in this digital landscape. Cybersecurity is not just about technology; it's about understanding risks, anticipating threats, and fostering a security-focused culture. As threats evolve, so must our strategies to counter them.

Reflect and Apply Questions

1. **AI in Daily Life:** Consider how AI is used in everyday technology like smartphones. How do you think similar AI technology can be used to improve cybersecurity?
2. **Identifying Your Cybersecurity Interest:** Reflect on the cybersecurity topics covered in this chapter. Which area interests you the most, and what basic skills would you like to develop in this area?
3. **Promoting Security Awareness:** Think of ways you could improve your own cybersecurity practices at home or work. What steps would you take to enhance your personal digital security?
4. **Ethical Considerations in Technology Use:** Recall a time when you had to ensure your actions online were ethical and legal. How does this experience relate to the principles of ethical hacking?
5. **Emerging Technology and Security:** Choose a new technology you've recently learned about. What potential cybersecurity risks do you think it might have, and how could these be addressed?

Chapter 10: Future-Proofing Your IT Career

In an era where technology evolves at an unprecedented pace, future-proofing your career in IT is not just advantageous; it's a requirement. Chapter 10 discusses strategies and insights necessary for staying relevant in the endlessly advancing field of Information Technology. From adapting to technological changes to exploring the nuances of specialization versus generalization, this chapter provides a roadmap for enduring success in IT. It emphasizes the importance of continuous learning and education, fostering innovation and creativity, and the value of community involvement. Understanding these elements is key to surviving and thriving in the industry.

Adapting to Technological Change

Technological evolution in the IT industry is a constant and rapid process, bringing forth new advancements and innovations at an unprecedented rate. This progression is driven by the continuous emergence of new technologies, such as AI, IoT, cloud computing, and more. Each of these technologies brings about significant shifts in how IT specialists work, the tools they use, and the strategies they employ. It's important for those in IT to not only recognize these trends but also understand their implications. This understanding helps professionals anticipate future changes, adapt their skills accordingly, and stay relevant in their field. As technology continues to advance, the ability to evolve with it becomes an indispensable aspect of maintaining a successful IT career.

Balancing Depth and Breadth of Knowledge: This concept involves mastering core areas of expertise while also having a broad understanding of various technologies and trends. Deep expertise in a specific area, such as network security or database management, provides a strong foundation and makes one highly valuable in that niche. At the same time, a broad understanding of various IT disciplines helps in adapting to different roles and collaborating effectively with cross-functional teams. This balance ensures that professionals are not only experts in their field but also versatile and adaptable to the changing technological landscape.

Real-World Example: Sarah had been working as a Network Security Analyst for several years at a mid-sized tech company. She was deeply skilled in her area, known for her expertise in securing network infrastructures. However, understanding the volatile nature of the tech industry, Sarah had always been keen on expanding her skill set. She spent her free time learning about cloud technologies and data analytics, attending workshops, and completing online courses.

When her company underwent restructuring, the network security department faced significant downsizing. While many of her colleagues with specialized skills in the network security department were let go, Sarah's diverse skill set became her saving grace. Management recognized her versatility and instead of letting her go, they transferred her to the cloud solutions team. In her new role, Sarah applied her broad knowledge to enhance the company's cloud security posture. Her story is a testament to the importance of balancing depth with breadth in IT skills. It highlights how a broad skill set can provide job security and open up new career opportunities, even in challenging situations.

The Role of Continuous Learning: Continuous learning is a cornerstone of a successful career, especially given the rapid pace of technological advancements. It's not just about acquiring new technical skills; it also involves understanding evolving industry trends, methodologies, and best practices. Continuous learning can take many forms, from formal education like degrees and certifications to informal learning through online resources, workshops, and industry seminars. This commitment to lifelong learning ensures that IT professionals remain adaptable, versatile, and competitive in the field. Moreover, continuous learning fosters innovation and creativity, enabling professionals to contribute effectively to their organizations and the broader community. T

Impact on Career Paths: Technological advancements significantly reshape career paths in the IT industry. As new technologies emerge, they create novel roles and specialties, while evolving or phasing out others. This consistently shifting scene requires IT professionals to be agile in their career planning, often necessitating changes in specialization or the acquisition of new skills. For example, the rise of cloud computing has led to a surge in demand for cloud specialists, while roles focused on traditional data center management have seen a shift. Similarly, the growth of cybersecurity has led to a higher demand for specialists skilled in threat analysis and mitigation. Understanding these trends and being prepared to pivot or upskill as necessary is imperative for maintaining a relevant and prosperous career in IT.

Community Involvement: Engaging in forums, groups, and professional associations can offer opportunities for learning, networking, and staying up-to-date of industry developments. It allows experts to exchange ideas, seek advice, and gain insights from peers and industry leaders. Community involvement can also open doors to new job opportunities, collaborations, and mentorship. For newcomers and seasoned professionals alike, being part of a community fosters a sense of belonging and contributes substantially to professional development.

Individuals can seek opportunities by reaching out to businessses, schools, churches, etc. in their local community. Quite often, especially in smaller cities, these locations don't have the money or resources to have an IT person on staff. You can offer your services at no cost as a learning experience for yourself at the benefit of the business. This will help build relationships that could open doors in the future. This will also provide you with real-world hands-on experience that can be invaluable to talk about in an interview.

Anticipating Future IT Roles: Anticipating future roles requires an understanding of how technological trends are shaping the industry. For instance, the advancement in areas like quantum computing, augmented reality, and blockchain technology promises to open up new specialties. Professionals who keep an eye on these developments and prepare accordingly will be well-positioned to take advantage of these emerging opportunities. Adapting to future roles may involve acquiring new skills, retraining, or even shifting to entirely new domains within Information Technology.

Continuous Learning and Education

In the IT industry, it is imperative to identify areas for continuous learning. This involves staying aware of industry trends and emerging technologies that are reshaping the field. IT experts should regularly assess their own skill sets against industry demands, identifying areas where they can improve or learn something new. This proactive approach ensures that they remain competitive and can adapt to changes in the industry. It is also beneficial for making informed decisions about which new technologies or methodologies to focus on, thereby maximizing the impact of their learning efforts.

Self-Guided Learning in IT: This offers flexibility and allows individuals to tailor their learning to specific interests and goals. This approach allows professionals to focus on specific skills or technologies that are most relevant to their career goals. With the abundance of online resources like tutorials, webinars, courses, and documentation, self-guided learning is both accessible and practical. It empowers IT professionals to stay current with the latest developments and innovations in the field. Additionally, this type of learning fosters a habit of independent research and problem-solving, skills that are valuable in the industry.

Leveraging Online Resources and Communities: These platforms offer access to a wealth of knowledge, including tutorials, forums, blogs, and online courses, covering a vast array of IT topics. Engaging with online communities, such as GitHub or Stack Overflow, can provide practical insights and real-world problem solving skills. These resources also offer opportunities for collaboration, networking, and peer support, which are valued for both learning and professional growth. By effectively leveraging these online platforms, IT professionals can stay up-to-date with current trends, learn new skills, and gain practical advice from peers in the industry.

Certifications Role in Professional Development: They are essential tools for IT pros to validate their skills and knowledge in specific areas, such as cybersecurity, cloud computing, or network management. Obtaining certifications can significantly boost career prospects, as they are often recognized and valued by employers globally. They also ensure that specialists are up-to-date with the latest technologies and practices in their field. Additionally, the process of preparing for certifications often involves comprehensive learning and practical application, further enhancing one's skill set. Thus, certifications are not just credentials but also key drivers of continuous learning and career growth in IT.

Integrating Learning into the Workday: This is important for individuals aiming to stay current and advance in their careers. Strategies for this include setting aside specific times for study, such as during breaks or commuting, and utilizing online resources or podcasts. On-the-job training, where new skills are applied in practical scenarios, is also an effective way to learn. It's advised to establish a habit of regular learning, whether through following online tutorials, engaging in webinars, or experimenting with new technologies. This proactive approach to learning ensures that professionals can keep pace with the rapid developments in the field, enhancing their skills and career prospects.

Specialization vs. Generalization

In the IT industry, 'specialization' refers to focusing on a specific technology or area, developing deep expertise in that niche. Specialists are known for their extensive knowledge and proficiency in a particular domain, such as cybersecurity, data analytics, or cloud computing. In contrast, 'generalization' involves having a broad skill set across various IT disciplines. Generalists are versatile, with a wide-ranging understanding that allows them to work across multiple aspects of the industry. They can adapt to different roles, from system administration to basic programming tasks. The choice between specialization and generalization depends on individual career goals, market demands, and personal interests. Both paths offer unique advantages and challenges in the developing world of technology.

Pros and Cons of Specialization: Specialists often become highly sought-after for their deep expertise and can command higher salaries. They are usually the first choice for roles that require in-depth knowledge and are well-positioned to lead projects in their area of expertise. However, specialization has its downsides. It can lead to a narrow focus, potentially limiting career flexibility. Specialists might find fewer job opportunities outside their niche, and rapid technological changes can render some specialized skills obsolete. Therefore, while specialization can lead to expertise and recognition, it also requires a commitment to staying current within a specific domain.

Pros and Cons of Generalization: Generalists can adapt to various roles within an organization, making them valuable for their versatility. They can bridge gaps between different departments and understand the broader picture of IT projects. This adaptability can lead to diverse career opportunities. However, being a generalist can also mean facing stiff competition in the job market, as they might not stand out in any one particular area. Additionally, while generalists have a broad understanding, they might lack the depth of knowledge required for highly specialized roles. Balancing breadth and depth can be a challenging aspect of following a generalist path.

Market Demand: This can play a decisive role in the decision between specialization and generalization. In certain periods, the market may favor specialists with deep knowledge in areas like AI or cybersecurity. At other times, organizations might prioritize generalists who can wear multiple hats, especially in smaller teams or startups. IT professionals should closely monitor industry trends and demands to make informed decisions about their career paths. Understanding which skills are in high demand can guide them in choosing whether to specialize or generalize, ensuring their skill set aligns with market needs.

Balancing Specialization with General Skills: Finding a balance between specialization and having general skills is key for a well-rounded IT career. This approach involves developing deep expertise in one area while maintaining a broad knowledge base in other IT disciplines. For instance, a network specialist might also cultivate a basic understanding of programming and database management. This balance enhances career flexibility, allowing experts to adapt to various roles while still offering specialized knowledge where it's most needed. This hybrid approach can provide the best of both worlds, offering security and versatility in the job market.

Navigating Career Paths: Piloting career paths requires careful consideration of one's interests, strengths, and market trends. Those leaning towards specialization should be prepared to invest time and resources in developing profound expertise in a chosen area. On the other hand, generalists should focus on acquiring a diverse skill set and staying adaptable. Individuals might start as generalists, gradually moving towards specialization as they identify areas they are passionate about. Alternatively, specialists might broaden their skill set to increase their marketability. Regular self-assessment and industry research help to make informed decisions about career direction in the tech field.

Specialization Case Study: John, a database administrator, dedicated his career to mastering SQL and database management. His deep expertise led to a role as a senior database architect, where he spearheaded major projects. John's specialization made him a go-to expert in his field, leading to recognition and lucrative consulting opportunities.

Generalization Case Study: Emma started her IT career with a broad skill set, working in roles from system analysis to basic coding. Her versatility allowed her to move into project management, where she leveraged her diverse experience. Emma's generalist background was important to her success in managing multidisciplinary teams and projects.

Innovation and Creativity in IT

In the IT industry, building a culture of innovation is key for staying competitive and driving progress. This involves creating an environment where new ideas are encouraged and valued. Organizations should foster a mindset of continuous improvement, where experimentation and learning from failures are seen as pathways to innovation. Encouraging cross-departmental collaboration can also spark creative solutions, as diverse perspectives often lead to groundbreaking ideas. Providing resources and time for research and development is another main factor. Leaders in IT should act as role models in innovative thinking, inspiring their teams to think creatively. Celebrating innovative achievements, no matter how small, can reinforce a culture where creativity is nurtured and rewarded.

The Role of Creativity in Problem Solving: Creativity enables individuals to think beyond conventional solutions and explore new possibilities. Creative problem-solving involves looking at challenges from different angles, questioning existing assumptions, and brainstorming novel ideas. This approach can lead to more effective and efficient solutions, as it opens up a wider range of possibilities. Encouraging team members to take creative risks and offering a safe space where all ideas are considered can lead to innovative solutions. Creative problem-solving is not just about having original ideas but also about being able to connect seemingly unrelated concepts in useful ways. This skill becomes increasingly important as problems become more complex and intertwined.

Leveraging Emerging Technologies for Innovation: Professionals can leverage these new tools to create solutions that were previously impossible. For instance, using AI and machine learning can lead to advanced data analysis and automation capabilities. Exploring the potential applications of blockchain beyond cryptocurrency, like in supply chain management or secure voting systems, can open new avenues for innovation. Adopting a mindset of exploration and curiosity about technologies like IoT, augmented reality, or quantum computing can lead to the development of cutting-edge solutions. Staying updated on technological advancements and understanding their potential applications is paramount to leveraging them for innovation.

Collaboration and Diverse Perspectives: Collaboration and the incorporation of diverse perspectives are fundamental for driving innovation. When individuals with different backgrounds, skills, and viewpoints come together, the potential for creative solutions increases significantly. Collaborative environments encourage the exchange of ideas and knowledge, leading to more comprehensive and well-rounded solutions. Embracing diversity in teams not only refers to cultural or gender diversity but also diversity in skills, experiences, and ways of thinking. Effective collaboration often involves open communication, mutual respect, and a shared vision. Creating opportunities for team members to collaborate on projects, brainstorm sessions, or hackathons can be highly beneficial for nurturing innovation.

Personal Development for Fostering Creativity: This involves continuously nurturing one's creative thinking skills through various means. Engaging in activities outside of work that stimulate creativity, such as arts, music, or even creative writing, can be beneficial. IT professionals should also seek to expand their knowledge base beyond their immediate field, as cross-disciplinary learning can inspire innovative ideas. Attending work shops on creative thinking, practicing brainstorming techniques, and maintaining a curious and open-minded attitude are all important. Personal development in creativity is not a one-time effort but an ongoing process that enriches one's professional and personal life.

Case Study of Innovative IT Solutions: A notable example of innovation in IT is the development of the Smart Traffic Management System by a team led by Maya, a software engineer. Faced with increasing traffic congestion in her city, Maya utilized her expertise in IoT and data analytics to devise a solution. Her team developed an interconnected system of sensors and cameras that collected real-time traffic data. This data was then analyzed using AI algorithms to optimize traffic light timings and reduce congestion. Maya's solution eased traffic flow and reduced environmental impact from idling vehicles. This case study demonstrates how innovative thinking, combined with IT skills, can lead to practical solutions with a significant societal impact.

Career Flexibility and Diversification

The tech world is marked by rapid technological advancements and shifting market needs. For IT professionals, this means that career flexibility and the ability to diversify skills are more important than ever. Adapting to these changes requires an openness to learning and a willingness to step out of one's comfort zone. IT experts need to be agile, able to pivot between different technologies and roles as needed. Understanding this landscape involves staying informed about industry trends, emerging technologies, and understanding how they can impact various roles. Being adaptable in this ever-changing environment can lead to a more resilient and sustainable career.

Benefits of Career Diversification in IT: Diversifying one's career can open up a myriad of opportunities and provide a safeguard against the rapid changes in the technology sector. Individuals who have a varied skill set are often more attractive to employers as they can contribute in multiple capacities. Diversification can lead to greater job security, as reliance on a single skill or technology can be risky in this field. It also offers the chance to explore different areas of interest within IT, keeping the career journey exciting and fulfilling. Additionally, diversified skills can lead to a broader understanding of Information Technology as a whole, fostering a more integrated approach to tech solutions.

Strategies for Skill Diversification: Developing a diversified skill set involves strategic planning and continuous learning. Individuals should look for opportunities to cross-train in different domains, either through education or on-the-job experiences. This might include branching out from a core specialty to learn related technologies or skills. Engaging in side projects or freelance work can also provide practical experience in new areas. Networking with professionals from diverse IT backgrounds can offer insights into different fields and potential learning paths. It's also necessary to stay updated with industry trends to identify which skills are becoming increasingly relevant.

Industry Shifts and Trends: To uphold a flexible and diversified career, it is recommended to remain adaptable to industry shifts and trends. This requires a proactive approach to learning and staying informed about the latest developments in technology. IT pros should regularly review industry publications, attend webinars and conferences, and participate in relevant online communities. Understanding these trends can help in anticipating the direction of the market and identifying areas for skill development. Being adaptable to new technologies and methodologies ensures that individuals remain valuable assets to their organizations and the industry at large.

Transitioning Between Roles: Identify transferable skills that can be applied to new roles and actively seek opportunities for growth and development. This may require additional training or certification in new areas of expertise. Effective transition also involves networking, both to learn about new opportunities and to gain insights into different roles. One should have a willingness to take on new challenges, even when they fall outside their comfort zone. Successfully transitioning between roles can lead to a more dynamic and rewarding career.

Case Study: Linda began her career as a software developer. Over time, she developed an interest in data analytics. Recognizing the growing importance of this field, she pursued additional training and certifications. Her transition was a calculated move that combined her programming skills with new knowledge in data analytics. This shift not only led to a more fulfilling role but also positioned her as a valuable asset in her organization, blending her software development background with emerging data trends. Linda's story illustrates the advantages of adapting to industry changes and the benefits of a diversified skill set.

Community Involvement

Community involvement in the industry offers many benefits, from professional growth to personal support. Engaging with a community fosters a sense of belonging and provides an avenue for sharing knowledge and experiences. It's an effective platform for networking, where professionals can connect with peers, mentors, and industry leaders. Communities also serve as excellent resources for staying informed about the latest technologies and industry trends. Active participation in IT communities, whether online or in-person, can lead to new opportunities, insights, and collabor-ations that are invaluable for career advancement.

Types of IT Communities: The IT industry boasts a diverse array of communities, each catering to different interests and professional needs. Online forums and platforms like Stack Overflow or GitHub offer spaces for problem-solving and project collaboration. Local meetups and user groups provide opportunities for face-to-face networking and knowledge sharing. Professional associations, like the IEEE Computer Society or CompTIA, offer structured memberships with access to resources and events. Additionally, contributing to open-source projects can be a way to engage with a global community of developers. Each type of community offers unique benefits, and IT experts can choose to engage with one or several based on their interests.

Contributing to Communities: Active contribution to these communities can be very rewarding. Sharing expertise, whether through answering questions in forums, presenting at meetups, or contributing code to open-source projects, helps others and enhances one's own knowledge. Volunteering for community events or taking on a leadership role in organizing activities can be rewarding. These contributions help in building a professional reputation and can often lead to unexpected career opportunities. They also foster a sense of achievement and satisfaction from giving back to the community.

Building Professional Relationships: Communities in the IT industry play a huge role in building professional relationships. These networks can lead to lifelong connections with peers, mentors, and industry leaders. Active participation in communities often results in strong professional bonds based on shared interests and collaborative experiences. These relationships can be priceless when seeking career advice, job opportunities, or partners for new ventures. They also provide a support system that can be utilized during career transitions or challenging projects.

Community Involvement as a Career Strategy: Being active in groups is a strategic move for career development. It enhances one's visibility in the trade, helping to build a personal brand and reputation. Regular community involvement can also signify a commitment to the field and continuous learning, traits that are highly valued by employers. It provides a platform to demonstrate skills and knowledge, potentially leading to job offers or collaborations. Moreover, group events and projects can be added to a portfolio, showcasing one's expertise and engagement with the industry.

Anticipating Future IT Roles

Emerging technologies such as AI, blockchain, and quantum computing are rapidly reshaping the IT landscape. These advancements are not just creating new tools and platforms; they are forging entirely new roles. For instance, the rise of AI is paving the way for roles focused on machine learning, AI ethics, and AI system integration. Blockchain technology is creating demand for blockchain architects and developers. Similarly, quantum computing is expected to open up new frontiers, requiring expertise in quantum algorithms and quantum hardware. Understanding these technologies and their potential applications is essential for individuals looking to stay ahead in their careers. Some additional insights are below:

1. **Artificial Intelligence and Machine Learning:**
    - IT roles, particularly in data science and analytics, are increasingly requiring skills in AI and ML. These technologies are transforming how businesses analyze data, automate processes, and make decisions, leading to a need for experts skilled in these areas.
2. **Internet of Things (IoT):**
    - The proliferation of IoT devices impacts network management and security roles, as IT specialists must ensure these devices are secure and efficiently integrated into existing networks.
3. **Blockchain:**
    - Blockchain technology is not just for cryptocurrencies but also for secure transactions and data management. IT professionals in cybersecurity, database management, and systems architecture need to understand blockchain to implement secure, decentralized systems.
4. **Cybersecurity:**
    - The repeatedly changing threat landscape makes cybersecurity a continuously relevant field. IT pros, especially in network security, incident response, and ethical hacking, need to stay abreast of the latest threats and defense mechanisms.

Skillsets for Future Jobs: The skillsets required for future jobs will likely be a blend of advanced technical knowledge and soft skills. Technical skills in areas like cybersecurity, data science, and cloud computing will remain in high demand, but they will need to be complemented with skills in emerging technologies. Soft skills such as adaptability, problem-solving, and effective communication will be equally important, especially as IT roles become more integrated with business operations. Additionally, skills in project management, leadership, and innovation will be vital for navigating the fast-paced changes and cross-functional nature of future roles.

Impact of Remote Work and Globalization: Remote work and globalization are significantly influencing the future of tech roles. The rise of remote work is creating opportunities for industry professionals to work for global companies without geographical constraints. This shift is likely to increase demand for skills in cloud computing, virtual collaboration tools, and cybersecurity for remote work environments. Globalization also means that individuals need to be aware of different cultural contexts and regulatory environments. As these teams become more distributed, the ability to work effectively across different time zones and cultures will become increasingly important.

Innovation and Entrepreneurial Opportunities: As technology continues to advance, there will be a growing need for experts who can develop innovative solutions to new challenges. This might involve designing new software, developing new methods for data analysis, or creating more efficient systems. The tech industry's fast-paced environment provides fertile ground for entrepreneurial ventures, particularly in emerging fields like AI and IoT. IT professionals with an entrepreneurial mindset, who can identify market needs and develop technological solutions, will find numerous opportunities to innovate and lead in their field.

As we conclude this exploration of future-proofing your IT career, it's clear that the key to longevity and success in this dynamic field lies in adaptability and continuous learning. The scene is constantly evolving, presenting new challenges and opportunities. Embracing these changes, whether through specialization in emerging technologies or maintaining a versatile skill set, is important. Networking, community involvement, and staying informed about industry trends play significant roles in navigating your career path effectively. The future of IT is not just about technical prowess but also about the ability to innovate, collaborate, and adapt in technological environments.

Reflect and Apply Questions

1. **Specialization vs. Generalization:** Reflect on your own interests and strengths. Would you prefer to specialize in a specific area of IT or develop a broad range of skills? Why?
2. **Emerging Technologies:** Think of an emerging technology you find fascinating. How could you incorporate learning about this technology into your career development plan?
3. **Adapting to Change:** Recall a time when you had to adapt to a significant change. How did you handle it and what did you learn that could apply to adapting in the IT industry?

4. **Continuous Learning:** What strategies would you use to incorporate continuous learning into your routine, ensuring your IT skills remain current and relevant?
5. **Community Involvement:** How could participating in an IT community benefit your career development, especially as someone new to the field?
6. **Career Flexibility:** Imagine a scenario where your current IT role is evolving. How would you prepare to adapt to this change and what new skills would you need to develop?

Chapter 11: The New Era of IT: Remote Work and Methods to Stay Ahead

In this chapter we explore the transformative effects of remote work on the IT industry and the skills necessary to navigate this evolving field. As the traditional office setting gives way to more flexible, distributed work environments, individuals face both opportunities and challenges. We will examine the impact of remote work on IT operations, the balance between its advantages and drawbacks, and the skills needed for success in remote settings. It also explores future trends in remote work, including the influence of emerging technologies like AI, VR, and AR, and the evolving cybersecurity landscape. The aim is to equip you with the knowledge and tools to thrive in this new era, where adaptability and continuous learning are paramount.

Remote Work's Impact on IT Operations

The shift to remote work has drastically altered IT infrastructure needs. Companies have increasingly turned to cloud services to provide flexible, scalable solutions that can be accessed remotely. This transition requires a reevaluation of network capacities and data storage solutions, with a greater emphasis on cloud-based platforms like AWS, Microsoft Azure, or Google Cloud. VPNs have become essential for secure remote access, demanding robust implementations and management. This shift has also led to a change in budget allocations, with increased investments in cloud services and cybersecurity measures to safeguard remote operations. The challenge for IT operations is to ensure that these infrastructures are not only efficient but also secure and resilient.

Cybersecurity Focus: The shift towards remote work has underscored the importance of robust cybersecurity measures. This new normal introduces vulnerabilities, such as less secure home Wi-Fi networks and a heightened chance of phishing attempts. To navigate these challenges, IT departments are enhancing their defenses with sophisticated endpoint protection, multi-factor authentication, and comprehensive security evaluations. Equally vital is educating staff on security awareness, given that human oversight can significantly contribute to breaches. The objective is crafting a secure telecommuting framework that safeguards organizational and personal data alike, while maintaining operational efficiency.

Communication and Collaboration Tools: The expansion of remote work has necessitated the integration of advanced communication and collaboration tools into IT operations. Tools like Slack, Microsoft Teams, and Zoom have become common, facilitating real-time communication and collaboration. Organizations must ensure these tools are seamlessly integrated, user friendly, and secure. This involves not only technical implementation but also addressing challenges related to bandwidth, data privacy, and user training. The effectiveness of these tools underpins productivity and strengthens team dynamics, even when working remotely.

Support and Maintenance Challenges: Remote work poses unique challenges for tech support and maintenance. Traditional in-person support has shifted to remote assistance, requiring IT staff to adapt their strategies and tools. Remote troubleshooting, virtual helpdesks, and remote desktop tools have become standard practices. However, issues like hardware failures still pose big challenges, as they often require physical intervention. Organizations must develop efficient protocols for handling such situations, possibly including partnerships with local repair services or implementing a robust hardware replacement policy.

Remote IT Workforce Management: This includes implementing tools for tracking productivity and work hours, while also ensuring employees are not overworked. Regular virtual meetings and check-ins are important for maintaining team cohesion and ensuring that everyone is aligned with their respective goals and responsibilities. Managers must also focus on fostering a positive remote work culture that supports employee wellbeing and encourages open communication. Successful remote workforce management requires balancing supervision with trust.

Pros and Cons of Remote Work

Increased Flexibility and Work-Life Balance: One of the most popular benefits of remote work is the increased flexibility it offers. This flexibility allows employees to create schedules that fit their personal lives and responsibilities better. It often leads to an improved work-life balance, as individuals can manage their time more effectively, blending work and personal activities in a way that suits them best. This adaptability can result in increased job satisfaction and reduced stress. However, it requires good time management skills to avoid work encroaching on personal time. For many, the ability to work from anywhere and avoid a daily commute adds hours back into their day, which can be used for personal development, family, or leisure activities.

Challenges in Maintaining Boundaries: While remote work offers greater flexibility, it also presents challenges in maintaining clear work-life boundaries. The absence of a physical separation between work and personal space can lead to the blurring of these boundaries, potentially resulting in longer work hours and difficulty in 'switching off' from work mode. This can lead to burnout and stress, as the home environment becomes an extended workplace. Employees may find themselves working irregular hours or being unable to fully disengage from work responsibilities. Establishing a dedicated workspace and setting firm working hours are crucial in mitigating these challenges. It's important for both employers and employees to recognize and address these boundary issues to maintain a healthy work-life balance.

Reduction in Commuting Stress: Eliminating the daily commute is a clear advantage of remote work. Commuting can be a major source of stress and lost time for many employees, and its reduction can lead to improved well-being and productivity. The time saved from commuting can be utilized for personal interests, additional rest, or spending time with family. This reduction in travel can also lead to financial savings and a decrease in environmental impact due to less transportation use. However, the loss of a structured commute means employees must find new ways to outline the beginning and end of their workday, which is important for mental health and maintaining a routine.

Potential Isolation and Communication Barriers: The shift to remote work can lead to feelings of isolation, particularly among those who value workplace social interactions. The absence of in-person conversations might also introduce obstacles, risking misunderstandings or a feeling of separation from the team. This situation can affect an employee's connection to their colleagues and their happiness at work. To mitigate this, remote teams should prioritize regular online gatherings and socializing opportunities. Encouraging digital coffee breaks or team events helps in nurturing community bonds. Adopting reliable communication tools and practices ensures everyone feels included and an active part of the company's culture.

Impact on Physical and Mental Health: Remote work has impacts on both physical and mental health. On the positive side, it can reduce stress related to commuting and office politics and offer a more comfortable and personalized work environment. However, it also presents challenges such as the risk of a sedentary lifestyle due to less physical movement throughout the day. Additionally, the lack of separation between work and home life can lead to increased stress and mental health issues if not properly managed. It's advised that remote workers establish routines that include regular exercise, take breaks away from their workstations, and create a workspace that is ergonomically sound to support their physical health.

Cost Savings vs. Home Office Expenses: Working remotely can lead to cost savings for both employers and employees. Employees save on commuting costs, work attire, and often food expenses. However, these savings can be offset by the need to invest in a suitable home office setup, including ergonomic furniture, technology, and increased utility bills. While some employers provide stipends for home office equipment, others do not, placing the financial burden on the employees. Balancing these costs requires careful consideration and planning. Employers offering remote work options should consider supporting their employees in setting up an efficient and ergonomic home office to ensure their comfort and productivity.

Skills for Remote IT Professionals

In remote IT roles, certain technical competencies become more prominent. Mastery in cloud computing stands out, as numerous remote functions depend on cloud solutions for storing data and hosting applications. A thorough grasp of cybersecurity becomes indispensable with the heightened vulnerabilities of remote connections. Skills in managing networks, particularly in diagnosing and enhancing network efficiency from afar, receive great appreciation. Knowledge of virtualization technologies proves useful for crafting adaptable IT setups. Also, being skilled in employing diagnostic tools for remote issue resolution is helpful for keeping operations smooth.

Effective Digital Communication: Honing digital communication skills is a keystone for individuals in remote roles. Clear and concise messaging in emails, chat platforms, and documentation helps prevent misunderstandings and project delays. Proficiency with virtual meeting software and adherence to online communication norms are essential for replacing in-person meetings. The art of articulating ideas effectively in virtual settings ensures that contributions are recognized and team cohesion is maintained. Additionally, the ability to communicate complex technical information in an understandable manner to non-IT colleagues fosters better understanding and collaboration in a distributed team setup.

Time Management and Self-discipline: Without the structured environment of an office, remote workers must be adept at managing their time effectively. This includes prioritizing tasks, setting realistic deadlines, and avoiding distractions common with home environments. Good self-discipline helps in maintaining a consistent work schedule, ensuring that work responsibilities are met efficiently. It's also important to take regular breaks to avoid burnout, as the lines between work and personal life can often blur in a remote setting. Time management tools and techniques, like the Pomodoro Technique or time-blocking, can be particularly helpful for remote workers.

Remote Collaboration Tools: Tools like Slack and Microsoft Teams facilitate communication and project management in distributed teams. Proficiency in these tools involves not just understanding their technical aspects but also knowing the best practices for their use, such as appropriate messaging etiquette and efficient task tracking. Being able to troubleshoot common issues with these tools will help ensure smooth team operations. Remote IT workers should be comfortable switching between different tools and platforms, as project needs and team prefer-ences can vary widely.

Remote Troubleshooting and Support: Individuals must be able to diagnose and resolve issues without being physically present, often relying on verbal or written descriptions of the problem. This demands a high level of technical knowledge, excellent communication skills, and patience. Remote users should also have familiarity with remote desktop tools and diagnostic software. Being able to guide non-technical users through troubleshooting steps in a clear and calm manner is valued. Remote workers should be prepared for a wide range of issues, from simple connectivity problems to complex system failures.

Maintaining Security and Privacy: IT experts need to be well-versed in best practices for securing remote networks, such as implementing strong password policies and using secure Wi-Fi connections. Knowledge of encryption tools and VPNs are important for protecting data transmission. Remote experts must also be aware of privacy laws and regulations, ensuring that their practices comply with these standards. Regular updates including patching software and systems is required to protect against vulnerabilities.

Continuous Learning and Adaptability: The tech industry is constantly evolving, and remote workers need to keep up with new technologies and methodologies. This might involve self-study, online courses, or virtual training. Adaptability is also about being open to changing work processes and tools, as remote work environments can rapidly evolve. Welcoming feedback and the ability to adapt in changing one's tactics are invaluable in a remote work context. Continuous learning and adaptability enhance an individual's skill set and contributes to the overall resilience and agility of the remote team.

Future Trends in Remote IT Work

AI and Automation: These technologies promise to automate routine tasks, allowing IT experts to focus on more complex and strategic activities. AI-driven tools could enhance remote monitoring and management of IT systems, making maintenance more efficient. However, this shift also raises concerns about job displacement and the need for upskilling. IT workers will need to adapt by developing skills in AI and machine learning to stay relevant. The future of remote work will likely involve a synergy between human expertise and AI capabilities, optimizing operational efficiency and innovation.

Advancements in Virtual and Augmented Reality: These technologies can create immersive and interactive environments for remote collaboration, making virtual meetings more engaging and productive. They also have the potential to revolutionize remote training and support, providing professionals with more hands-on, realistic experiences. For instance, AR can be used for remote assistance, where IT support can guide users through complex tasks with overlay instructions. As these technologies mature, they could become standard tools for remote teams, bridging the gap created by physical distance.

Shifts in Cybersecurity Dynamics: The shift towards more remote work will inevitably change the cybersecurity scene. With a distributed workforce, the perimeter of security expands, making traditional security models less effective. This change will require new approaches, such as zero-trust security models and enhanced endpoint security. IT experts will need to be more vigilant about potential threats like phishing attacks, which can be more prevalent in remote settings. Remote workers should seek continuous education on cybersecurity best practices. Future remote work will likely see increased investment in cybersecurity tools and training, ensuring that remote environments are as secure as traditional office settings.

Evolving Communication Technologies: The future of remote work will be heavily influenced by evolving communication technologies. Upcoming advancements in these technologies aim to make remote collaboration more seamless and intuitive. We can expect developments in high-quality video conferencing, real-time language translation, and more sophisticated collaboration platforms that integrate various aspects of project management. These technologies will facilitate smoother communication and foster a more inclusive and global work environment. The challenge will be to ensure these technologies are accessible and easy to use, enhancing rather than complicating the remote work experience.

Sustainability and Environmental Impact: Reduced commuting and lower office energy consumption can contribute to a smaller carbon footprint. Additionally, the digitalization of many work processes reduces the need for physical resources, further supporting environmental sustainability. However, there are trade-offs, such as the energy demands of maintaining digital infrastructures. In the future, we may see more emphasis on green computing and efforts to minimize the environmental impact of remote operations, aligning technological advancement with ecological responsibility.

Resources to Stay Informed

In today's digital age, staying aware of the latest trends and technologies is paramount for IT professionals. RSS feeds, standing for "Really Simple Syndication," are a powerful tool for collecting online content. They allow users to receive updates from their favorite websites or blogs in a consolidated format. RSS feeds work by pulling the latest content from various sources into a single feed reader, providing a streamlined way to stay informed. By setting up RSS feeds, you can customize the influx of information to match your specific interests, ensuring you're always informed without being overwhelmed. This is particularly useful for experts who need to keep up with rapidly changing technologies and industry news. Feedly is a popular RSS feed app that is utilized by many IT pros.

Setting Up and Customizing RSS Feeds: To begin, one needs to choose an RSS feed reader, like Feedly, which acts as a central hub for all subscribed feeds. Once set up, users can start subscribing to various websites and blogs relevant to their IT interests. Customization is key; feeds can be organized into categories or lists, making it easier to navigate and prioritize reading. This personalized approach ensures that you receive the most relevant and current information in your field of choice.

Best Practices: To effectively utilize RSS feeds, it's important to adopt certain best practices. Regularly curating and updating the list of subscribed feeds ensures the information remains relevant and valuable. Categorizing feeds into topics or themes can help in efficiently navigating through large amounts of content. It's also beneficial to integrate RSS feeds with other tools, like using an automation tool to send updates to email or a Slack channel. Periodically reviewing and pruning the feed list helps in avoiding information overload and maintaining a high-quality stream of content.

Combining Multiple Information Sources: For comprehensive coverage, one should consider combining RSS feeds with other information sources. This might include subscribing to industry-specific newsletters, following tech influencers on social media, and listening to relevant podcasts. Each source offers a different format and perspective, providing a more wholistic understanding of industry trends and updates. By integrating these various sources, individuals can create a rich and diverse information ecosystem that keeps them well-informed and ahead in their field.

RSS Feeds and Continuous Learning: RSS feeds are an excellent tool for continuous learning in the IT sector. They provide a constant stream of the latest articles, tutorials, research papers, and news updates, making it easier to stay abreast of new technologies and methodologies. By dedicating time to regularly review RSS feeds, IT pros can ensure they are continuously learning and keeping up with industry advancements. This habit of staying informed also inspires new ideas and innovation in their work.

Balancing Your Information Diet: With the vast amount of content available, it's easy to fall into the trap of information overload. To maintain a healthy balance, be selective about your sources. Prioritize quality over quantity by choosing feeds, newsletters, and podcasts that consistently deliver value and relevance to your field. Regularly review and refine your subscriptions to discard those that no longer serve your professional growth or interests.

Implementing a schedule for consuming different types of content can also prevent burnout. Allocate specific times for reading articles, listening to podcasts, and engaging with communities. This not only helps in managing your time efficiently but also ensures that you're absorbing and reflecting on the information you receive.

Following Tech Influencers and Thought Leaders

Identifying the right tech influencers and thought leaders is key to gaining valuable insights. Start by defining your areas of interest or the specific technologies you want to learn more about. Look for influencers who not only have expertise but also actively engage with their audience and provide up-to-date, relevant content. Platforms like X(Twitter), LinkedIn, and YouTube are great places to start searching. Check their credentials, the quality of their discussions, and their follower engagement. Once you find influencers that align with your interests, their content can become a valuable resource for industry trends, new technologies, and innovative ideas.

Leveraging Social-Media: These platforms offer direct access to the thoughts and insights of industry leaders. Following influencers on these platforms can provide a steady stream of valuable information, from the latest tech news to deep dives into specific technologies. Engaging with their content through likes, comments, or shares can also open up opportunities for dialogue and networking. Additionally, many share links to comprehensive articles, research papers, or tutorials, providing a richer learning experience. It's important to curate your social feeds to ensure they are aligned with your professional learning goals.

Engaging with Online Tech Communities: Influencers often foster communities around specific technologies or topics within IT. By actively participating in these communities, you can deepen your understanding, ask questions, and share your own insights. Engaging in discussions sparked by influencer posts can lead to valuable connections and a more nuanced understanding of complex IT subjects. It's also a way to stay ahead of differing perspectives and emerging debates in the tech world. This active engagement not only broadens your knowledge but also increases your visibility within professional circles.

Critical Analysis of Shared Information: While tech influencers can provide valuable insights, it's important to critically analyze the information they share. Not all content may be accurate or unbiased, so it's important to verify facts and consider different viewpoints. Cross-reference information with credible sources and be cautious of content that seems overly promotional or one-sided. Developing the skill to discern quality content from misinform-ation is vital in the age of digital information. This approach ensures that the insights and knowledge gained from influencers are sound and can be reliably applied in your professional life.

Networking and Professional Growth: Following and interacting with tech influencers can significantly contribute to networking and professional growth. Engaging with their content, asking thoughtful questions, and participating in discussions can increase your visibility and establish your reputation in the field. Many influencers are well-connected industry professionals, and building a rapport with them can lead to mentorship opportunities and career advice. This interaction can also lead to connections with other experts in the field, expanding your professional network. In the digital age, building a professional network online can be as impactful as traditional, in-person networking.

Attending Virtual Conferences and Webinars

For effective learning and networking, it is imperative to select the right virtual tech conferences and webinars. Start by identifying your current skill gaps or areas where you want to expand your knowledge. Look for events that are specific to these interests or that are known for their high-quality content and speakers. It's also beneficial to consider events organized by reputable industry bodies or leading tech companies. Reading reviews or seeking recommendations from colleagues can provide insights into the value of these events. Remember, the goal is to choose events that not only align with your professional development goals but also offer opportunities for meaningful engagement.

Maximizing the Benefits of Virtual Attendance: Before the event, review the agenda and mark sessions of interest. During the event, engage actively by asking questions and participating in discussions. Many virtual events offer networking opportunities, such as breakout rooms or chat features, which can be great for connecting with peers and speakers. After the event, review your notes and reflect on how the new information can be applied to your work. Following up with contacts made during the event can also help in building your network.

Staying Updated: Attending virtual tech conferences and webinars is an excellent way to stay updated with the latest industry trends. These events often feature discussions on cutting-edge technologies and future industry directions. They provide insights from leading experts and practitioners in the field, offering a broader perspective on where the industry is headed. Staying informed about these trends is indispensable for IT specialists looking to future-proof their careers. Understanding current trends can help in making informed decisions about areas for skill development and career advancement.

Virtual Environment Networking: This practice, though sometimes challenging, can be highly effective if approached strategically. During virtual events, take advantage of features like chat rooms, Q&A sessions, and virtual networking lounges. Introduce yourself clearly and concisely, and don't hesitate to reach out to speakers or participants who share your interests. Follow-up is key; connect with new contacts on social networks like LinkedIn and continue the conversation. The goal is to build meaningful connections that can provide mutual benefits over time.

Applying Learnings to Your Career: This is an important step to make the most of these experiences. After attending an event, take the time to reflect on the main takeaways and how they can be integrated into your current role or projects. Share new insights with your team or superiors, which can also help in establishing your role as a knowledgeable and proactive team member. Implementing new techniques or strategies learned from these events can lead to improved efficiency, innovation, and problem-solving in your work.

Tech Podcasts and Vlogs

Selecting quality tech podcasts and vlogs is necessary for effective learning. Start by identifying the topics you're most interested in, whether it's general IT, specific technologies, or industry trends. Look for content creators who possess the technical expertise and have a knack for engaging and clear communication. Recommendations from colleagues or online forums can be valuable. Also, consider the frequency of updates and the podcast or vlog's audience engagement, as these can be indicators of the content's relevance and quality. The goal is to find resources that not only inform but also inspire and provoke thought.

Diverse Content: Exploring a range of podcasts and vlogs will help to gain a well-rounded perspective. Diverse content can range from technical tutorials and deep dives into specific technologies to discussions about career paths and industry trends. It's beneficial to mix technical content with material that focuses on soft skills, emerging technologies, or interviews with industry leaders. This variety ensures a comprehensive understanding of IT, encompassing both the hard skills of the trade and insights into the industry's broader context. Diversifying sources will also help in developing a more holistic view of the IT landscape.

Incorporating into Learning Routines: Dedicate specific times for this, such as during a commute or as part of a morning routine. It's important to not overload yourself; select a few podcasts or vlogs that align with your learning objectives and follow them regularly. You can rotate these choices periodically to cover different topics or to prevent content fatigue. Using tools that allow you to bookmark or save episodes for later reference can also be helpful. The aim is to complement your learning, not to replace formal training or hands-on experience.

Engaging with Content Creators: Many content creators appreciate feedback and questions from their audience, which can lead to more tailored content or deeper discussions. Commenting on episodes, participating in related online discussions, or even reaching out directly to the creators can provide additional insights. It's also a way to connect with a community of like-minded individuals who share your interests. By actively engaging, you not only enhance your own learning but also contribute to the broader IT community.

Leveraging News Aggregation Apps and Tools

News aggregation tools are digital platforms that compile news and updates from various sources into one centralized location. These tools are invaluable for IT experts who need to stay informed about the latest developments in a fast-paced industry. They pull content from multiple websites, blogs, and news services, providing a diverse range of information. Users can customize these feeds to focus on specific areas of IT, ensuring they receive news that's most relevant to their interests and needs. By using these tools, individuals can save time and streamline their information intake, ensuring they don't miss important updates.

Selecting the Right Tool: The tool should allow you to tailor your news feed to specific IT domains or topics of interest. The range of sources available is also important to ensure a broad and diverse information spectrum. Additionally, the user interface and ease of use play a big role in your daily interaction with the app. Some popular news aggregators offer features like keyword alerts or the ability to save articles for later reading, enhancing their utility. Feedly, Inoreader, and Flipboard are popular news aggregation tools.

Customizing Feeds for Relevant News: Start by identifying areas that you want to stay updated on, such as cloud computing, cybersecurity, or emerging technologies. Most aggregation tools allow you to set up filters or channels based on these topics, ensuring that you receive news that's directly relevant to your interests. While it's beneficial to have a broad overview of IT, diving deep into specific topics is also needed for specialized knowledge. To achieve this balance, you can set up separate feeds or categories – one for general IT news and others for in-depth coverage of specific topics. Periodically review and adjust these categories to align with your changing professional needs. This approach ensures you stay well-rounded in your knowledge while also developing expertise in particular areas of interest.

Integrating News into Daily Routines: This can be done without causing information overload. Set aside specific times for news consumption, such as during morning coffee or in between work tasks. Using features like bookmarking or "read later" can help manage articles of interest without disrupting your schedule. It's also helpful to limit notifications to only the most relevant updates to avoid constant interruptions. Regularly revisiting and adjusting your aggregation settings can keep the flow of news manageable and relevant.

Using Aggregated News for Career Advancement: Staying informed through aggregated news tools can significantly contribute to career advancement. Knowledge of the latest technologies, industry trends, and best practices will position you as a well-informed and up-to-date expert. In job interviews, being able to speak knowledgeably about recent tech developments or industry news can be impressive to potential employers, demonstrating your passion and commitment to the field. It shows that you are proactive in your development and can bring valuable insights to the team.

Real-World Example: John, an IT professional, used his knowledge from aggregated news sources to stand out in a job interview. He mentioned a recent breakthrough in AI that he read about, relating it to the company's ongoing projects. His ability to discuss current industry trends demonstrated his expertise and also his enthusiasm for staying at the forefront of technology. This approach left a strong impression on the interviewers and played a key role in his successful job application. This example illustrates how staying informed through news aggregation can open doors and create opportunities in your career.

As we conclude this chapter, we've explored the multi-layered impact of remote work on IT operations and the innovative ways specialists can stay informed and ahead in their field. From adapting to the unique demands of remote environments to leveraging modern tools like podcasts and news aggregators, the landscape of Information Technology is evolving rapidly. This chapter has highlighted the importance of effective communication, continuous learning, and strategic use of technology in shaping a successful remote career.

Reflect and Apply Questions

1. **Remote Work Environment:** If you were to set up a home office for remote IT work, what elements would you include to ensure productivity and efficiency?
2. **Evaluating Information Sources:** How would you assess the reliability and value of a tech podcast, vlog, or news aggregator in your learning process?
3. **Engaging with Tech Communities:** Can you think of ways to actively engage with an online tech community or discussion initiated by a tech influencer?
4. **Applying Tech News:** Imagine discussing a recent technological advancement in a job interview. How would you prepare and present your thoughts to make a strong impression?
5. **Balancing Work and Life:** What strategies would you employ to support a healthy work-life balance while working remotely?

Chapter 12: Mastering Time Management and Enhancing Soft Skills

In this chapter we will discuss the skills that go beyond technical expertise in the IT world. In this fast-paced field, managing time efficiently and excelling in interpersonal interactions are as important as technical acumen. We provide actionable strategies for managing time effectively, catering specifically to the demands faced by IT specialists. Additionally, we explore soft skills—effective communication, teamwork, problem-solving, and emotional intelligence—that are fundamental to success in the field. These competencies are indispensable for handling complex team interactions, mitigating stress, and cultivating a rewarding IT career.

Essential Skills for IT Professionals

This section joins various skills important for success in the industry. It encompasses essentials of time management, communication skills tailored for IT pros, emotional intelligence in the workplace, and techniques for resilience and stress management. These skills are invaluable for navigating the challenges and demands of a career in IT effectively.

Basics of Time Management in IT

Effective time management starts with the prioritization of tasks. It's vital to distinguish between urgent and important tasks, focusing first on those that have the greatest impact on your goals and projects. Techniques like the Eisenhower Matrix can be helpful in categorizing tasks based on their urgency and importance. Prioritization also involves being realistic about what can be achieved within given timeframes and understanding the scope and requirements of each task. Regularly reviewing and adjusting priorities based on changing circumstances is advised, as the dynamic nature of IT often leads to shifting priorities.

Effective Scheduling Techniques: This involves planning your day or week in advance, allocating specific time slots to tasks based on their priority and complexity. Utilizing calendar tools and scheduling software can aid in this process. It's imperative to build in buffers for unexpected issues or tasks that may take longer than anticipated. Time blocking, a technique where specific blocks of time are allocated to different tasks or activities, can be particularly effective in managing a busy workload.

Delegating and Outsourcing: These involve identifying tasks that can be handled by other team members or external resources. This not only helps in managing your workload but also in building a collaborative team environment. Effective delegation requires clear communication of expectations and timelines. It's also important to delegate tasks to the right people, considering their skills and workload.

Avoiding Burnout: It is important to recognize the signs of burnout, such as chronic stress, fatigue, and decreased productivity. Implementing strategies like taking regular breaks, setting boundaries between work and personal life, and engaging in activities outside of work can help maintain a healthy balance. Time management plays a significant role in preventing burnout by ensuring workloads are sustainable and deadlines are realistic.

Setting Realistic Goals: Goals should be specific, measurable, achievable, relevant, and time-bound (SMART). They provide direction and help in focusing efforts on what's most important. Rember to regularly review and adjust goals based on progress and any changes in circumstances. Setting realistic goals helps in maintaining motivation and prevents the feeling of being overwhelmed.

Managing Distractions: This is imperative in tech environments where multiple stimuli can lead to decreased productivity. Strategies include setting specific times for checking emails and messages, using tools to block distracting websites, and creating a work environment conducive to focus. It's also recommended to communicate with colleagues about your focused work periods to minimize interruptions. Regular breaks can help in maintaining concentration over longer periods.

Time Management in Agile Environments: Agile methodologies emphasize iterative work and frequent reassessment of priorities. This necessitates regular planning sessions to adjust to changing requirements and priorities. Techniques like daily standups and sprint planning can help in maintaining focus on current priorities while being adaptable to changes. Time management in agile environments is about balancing structure with flexibility.

Balancing Multiple Projects: This is common in IT and requires strategic time management. It involves understanding the scope, deadlines, and requirements of each project and allocating time accordingly. Communication with stakeholders and team members is necessary to ensure expectations are aligned. Techniques like kanban boards can help in visualizing and managing the progress of multiple projects simultaneously. Balancing multiple projects requires staying organized and being proactive in managing workload.

Communication Skills for IT Professionals

Effective verbal communication in IT requires the ability to articulate complex technical concepts in a way that's accessible to both technical and non-technical audiences. For more technical discussions, using precise language and industry terminology is preferred, while for non-technical stakeholders, simplifying concepts without losing the essence is crucial. It involves active listening to understand the audience's level of knowledge and tailoring the message accordingly. Practicing empathy and patience, especially when explaining technical details to those less familiar with IT, enhances understanding and collaboration. Storytelling techniques can be effective in making technical information more relatable. Regular practice and seeking feedback from peers can greatly improve these communication skills.

Writing Skills: Clear, concise, and professional writing is necessary for creating effective emails, reports, documentation, and even code comments. It involves organizing thoughts logically, using simple language for clarity, and avoiding jargon when addressing non-technical audiences. Attention to detail, such as grammar and spelling, reflects professionalism and ensures the message is well-received. Writing should be audience-specific, with the complexity of the content adjusted according to the reader's technical expertise. Additionally, developing the habit of proofreading before sending or publishing is vital to avoid misunderstandings or errors.

Active Listening and Feedback: Active listening helps foster better understanding and collaboration. It involves fully concentrating on the speaker, understanding their message, responding appropriately, and remembering the conversation. This skill is particularly important during team meetings, client discussions, and while gathering project requirements. Active listening also enhances the ability to provide and receive constructive feedback effectively. Encouraging open dialogue, asking clarifying questions, and showing empathy are the main components of active listening. This skill not only improves communication but also builds trust and respect within the team.

Non-Verbal Communication: In face-to-face meetings, maintaining eye contact, having an open posture, and nodding can convey engagement and understanding. In virtual settings, being mindful of your camera presence, such as maintaining eye contact with the camera, is recommended. Non-verbal cues can often convey more than words and can be used to reinforce or contradict what is being said. Being aware of your own non-verbal signals and interpreting others' correctly can enhance communication effectiveness.

Cross-Cultural Communication: In the globally connected IT world, cross-cultural communication skills are essential. Understanding and respecting cultural differences can lead to more effective collaboration and fewer misunderstandings. This includes being aware of different communication styles, negotiation techniques, and decision-making processes. Adapting your communication approach to suit different cultural contexts, such as being more direct or indirect, can improve interactions. Developing cultural competence involves continuous learning and exposure to diverse cultures, which can be achieved through training, travel, or working in international teams.

Conflict Resolution and Negotiation: Effective conflict resolution involves identifying the root cause of the conflict, actively listening to all parties, and working collaboratively to find a solution. It requires empathy, patience, and the ability to remain neutral. Negotiation skills are valuable when reaching agreements on project requirements, timelines, or resources. Successful negotiation involves clear communication, understanding the needs of all parties, and finding a mutually beneficial solution. These skills can resolve differences and foster a positive and productive work environment.

Communication in Remote and Hybrid Teams: Effective communication plays a big part in maintaining productivity and team cohesion. This involves regular check-ins and updates, using clear and concise language in written communications, and making the most of virtual meetings with clear agendas. Establishing communication norms, such as response time expectations and preferred communication channels, can enhance team coordination. Embracing digital collaboration tools and ensuring all team members are proficient in using them is also key. Building rapport through virtual team-building activities and informal interactions can strengthen relationships and improve communication in a remote setting.

Team Collaboration and Interpersonal Skills

Building Effective IT Teams: This requires a strong understanding of the project goals and the specific roles and skills needed to achieve them. Successful teams are characterized by a diversity of skills, perspectives, and problem-solving approaches. Open communication, trust, and shared objectives are foundational elements. Team-building activities and regular meetings can help strengthen relationships and enhance collaboration. It's also important to establish a supportive environment where team members feel valued and motivated to contribute their best.

Managing Team Dynamics: This requires an understanding of different personalities and working styles. It's important to recognize and value the diverse contributions of each team member. Effective team management involves facilitating collaboration, ensuring equitable work distribution, and addressing any interpersonal issues promptly. It's paramount to create an environment where team members feel comfortable sharing ideas and voicing concerns. Regular team-building exercises and open discussions about working styles can help in creating a cohesive team.

Collaboration Tools and Techniques: Effective collaboration in IT teams often relies on the use of various tools and techniques. Project management tools like Jira or Trello help in tracking tasks and progress, while communication tools like Slack or Microsoft Teams facilitate day-to-day interactions. It's recommended to choose the right tools for your team's needs and ensure all members are proficient in using them. Beyond tools, collaboration techniques like pair programming, brainstorming sessions, and regular stand-up meetings can enhance team productivity and engagement.

Conflict Management in Teams: This involves identifying sources of conflict early and addressing them constructively. Effective conflict resolution strategies include open communication, seeking to understand different viewpoints, and working towards mutually beneficial solutions. It's important for team leaders to mediate conflicts impartially and promote a culture of respect and understanding. Training in conflict resolution can also be helpful for team members to handle disagreements effectively.

Fostering a Culture of Inclusivity and Respect: This involves valuing diverse backgrounds, experiences, and perspectives. Inclusive practices include actively seeking input from all team members, ensuring equal opportunities for participation and advancement, and addressing any forms of bias or discrimination. Regular training and awareness programs can help in fostering an inclusive culture. Leaders play a key role in setting the tone and expectations for respectful and inclusive interactions.

Leadership and Influence: Leadership within teams is not just about formal roles but also about influence and the ability to inspire and guide others. Effective leaders are those who can motivate the team, set clear goals, and create a vision that others are excited to follow. They possess strong communication skills, demonstrate empathy, and are adept at problem-solving. Leadership also involves mentoring and developing team members, encouraging innovation, and supporting a positive team culture. Developing these leadership qualities can enhance one's impact within a team, regardless of official title.

Problem-Solving and Critical Thinking

Analytical Thinking involves breaking down complex problems into smaller, more manageable components to understand them better. This skill is indispensable in identifying the root causes of issues rather than just addressing symptoms. IT specialists use analytical thinking to evaluate different aspects of a problem, from technical details to the broader system implications. Being methodical and thorough in this analysis leads to more effective and sustainable solutions. Developing strong analytical skills often involves practicing with a variety of scenarios and learning from experienced professionals.

Strategies for Effective Problem-Solving: Techniques like root cause analysis help in identifying the underlying causes of problems. Decision matrices can aid in evaluating multiple solutions against a set of criteria. It's important to approach problems systematically, breaking them down into smaller parts and addressing each part in turn. Brainstorming sessions with team members can bring diverse perspectives and solutions. It is advised to document the problem-solving process and outcomes for future reference and learning.

Technology Selection: Critical thinking is encouraged when selecting technologies for projects. It involves evaluating the suitability of different technologies based on project requirements, scalability, and integration capabilities. IT professionals must consider both current needs and future growth, assessing the potential risks and benefits of each option. Critical thinking also involves questioning assumptions and looking beyond marketing hype to determine the true value of a technology. Peer reviews, pilot testing, and cost-benefit analysis are part of this evaluation process.

Creative Problem-Solving: This is increasingly important in Information Technology, where innovative solutions can provide a competitive edge. It involves thinking outside the traditional frameworks and experimenting with novel approaches. Creativity in IT isn't just about new ideas; it's also about recombining existing technologies in unique ways. To foster creativity, encourage a culture of experimentation and accepting that failure is a part of the innovation process. Brainstorming sessions, hackathons, and collaborative workshops can stimulate creative thinking.

Learning from Failure: This is a crucial aspect of problem-solving. It involves analyzing what went wrong, understanding the contributing factors, and using that knowledge to improve future efforts. A constructive approach to failure encourages a learning culture where risks can be taken without fear of blame. This perspective is vital for innovation and continuous improvement in the industry. It's important to document failures and the lessons learned from them, as they can be valuable for training and avoiding similar issues in the future.

Problem-Solving Mindset: Developing this mindset is about cultivating an attitude geared towards proactive resolution and improvement. This involves seeing problems as opportunities to learn and grow, rather than obstacles. It requires curiosity, a willingness to ask questions, and a commitment to continuous learning. It also involves being open to feedback and different approaches, constantly looking for ways to improve both personally and professionally. Encouraging a problem-solving mindset within an IT team can lead to more efficient and effective work processes.

Emotional Intelligence in the Workplace

Emotional Intelligence (EI) is the ability to recognize, understand, and manage our own emotions, and to recognize, understand, and influence the emotions of others. In the IT workplace, high EI contributes to better teamwork, improved problem-solving, and more effective communication. It's not just about being empathetic; it's also about self-awareness, self-regulation, and social skills. IT experts with high EI can navigate the stresses and complexities of the workplace more effectively. Understanding the basics of EI is the first step in applying these skills in everyday professional interactions.

Self-Awareness: This is an important component of EI, particularly for individuals who often work in high-stress environments. It involves understanding your own emotions, strengths, weaknesses, and values, and recognizing how they affect your actions and thoughts. Self-aware individuals are better equipped to handle workplace challenges and interactions. They can assess situations objectively and respond appropriately, rather than reacting impulsively. Developing self-awareness can involve reflection, seeking feedback, and being mindful of one's emotional responses in different situations.

Self-Regulation and Stress Management: In the context of EI, self-regulation refers to the ability to control and redirect disruptive emotions and impulses. In the IT industry, this skill is valuable for managing stress and maintaining professional conduct. It involves being able to stay calm and think clearly under pressure, which is crucial in fast-paced or crisis situations. It also includes being adaptable and maintaining a positive attitude, even in the face of adversity or rapid change. Techniques such as deep breathing, taking breaks, and positive self-talk can aid in self-regulation.

Empathy in Professional Interactions: Empathy is the ability to understand and share the feelings of others. In the workplace, it enables individuals to better connect with colleagues and clients, fostering a more collaborative and supportive environment. Empathetic communication involves active listening and being genuinely attentive to others' concerns and perspectives. It helps in building trust and resolving conflicts more effectively. Being empathetic doesn't mean agreeing with everyone; rather, it's about acknowledging and respecting different viewpoints.

Social Skills and Team Dynamics: Social skills in EI encompass a range of competencies from effective communication and conflict resolution to teamwork and leadership. In IT, where projects often require close collaboration, strong social skills are highly sought after. They enable professionals to build positive relationships, work effectively in teams, and influence others positively. Socially skilled experts can navigate workplace dynamics adeptly, contribute to a positive work culture, and lead by example. These skills can be developed through practice, feedback, and observing effective communicators and leaders.

Developing Emotional Intelligence: Developing EI involves a commitment to continuous learning and self-improvement. For IT professionals, this can mean seeking out resources and training, soliciting feedback from peers or supervisors, and engaging in self-reflection practices. Setting specific goals for improvement in areas like active listening, empathy, or stress management can provide direction. It's also beneficial to put oneself in challenging social situations that require the application of EI skills, as real-world practice is pivotal for development.

EI in Leadership and Management: Leaders with high EI can inspire and motivate their teams, manage conflict effectively, and create an environment where employees feel valued and understood. They are adept at understanding team dynamics and can navigate challenges with a balanced approach. Such leaders are typically approachable, open to feedback, and skilled in recognizing and utilizing the strengths of their team members. Cultivating EI in leadership roles can lead to improved team performance, higher job satisfaction, and better overall workplace harmony.

Resilience and Stress Management

Resilience is the ability to adapt well to stress, adversity, or change often encountered in a tech-focused workplace. This resilience is not just about enduring difficulties; it's about learning, growing, and emerging stronger from these challenges. In the IT sector, resilience might manifest as the capacity to handle high-pressure situations, tight deadlines, or rapidly evolving technologies without becoming overwhelmed. It involves maintaining a positive outlook, even in the face of setbacks or failures. Building resilience is a continuous process, involving self-awareness, stress management, and problem-solving skills. It's also about seeking balance, knowing when to push through a challenge and when to step back and recharge. Resilient IT pros are often more adaptable, efficient, and ultimately more successful in their careers.

Identifying and Managing Workplace Stressors: Common stressors include heavy workloads, tight deadlines, complex problem-solving, and the need to constantly update skills. Recognizing the signs of stress early, such as irritability, fatigue, or decreased productivity, is key to managing it effectively. Strategies for handling these stressors include time management, prioritizing tasks, and setting realistic goals. It's also important to establish boundaries between work and personal time. Open communication with supervisors about workload and expectations can help in finding solutions and managing stress. Additionally, developing a support network of colleagues and peers can provide a sounding board for sharing challenges and strategies.

Building Emotional Resilience: This involves developing the capacity to handle emotional challenges and bounce back from setbacks. It starts with self-awareness: understanding your emotional triggers and responses. Cultivating a positive mindset is paramount, focusing on solutions rather than dwelling on problems. Maintaining a sense of humor, practicing gratitude, and nurturing optimism are all part of building emotional resilience. Resilient individuals also learn from their experiences, viewing challenges as opportunities for growth rather than insurmountable obstacles. It's imperative to maintain a supportive network, as relationships with colleagues, friends, and family can provide emotional support and perspective. Regularly engaging in activities that promote emotional well-being, such as hobbies or relaxation techniques, can also bolster resilience.

Physical Health and Stress Management: Regular physical activity is one of the most effective ways to combat stress. It can enhance mood, improve sleep, and reduce anxiety. A balan-ced diet, rich in nutrients, also supports overall well-being and helps the body cope with stress. Adequate sleep is important, as it affects mood, energy levels, and the ability to handle stress. Establishing a regular sleep routine can greatly improve sleep quality. Additionally, avoiding excessive caffeine and alcohol, which can exacerbate stress and affect sleep, is recommended. Regular health check-ups can help identify and address any physical issues that might contribute to stress.

Mindfulness and Relaxation Techniques: These techniques are effective tools for reducing stress and enhancing overall well-being. Mindfulness involves being fully present in the moment and aware of your thoughts and feelings without judgment. Practices like meditation, deep breathing exercises, and yoga can help promote mindfulness. These techniques can lower stress levels, improve concentration, and enhance emotional clarity. Setting aside time daily for relaxation and mindfulness practice can create a significant positive impact on mental health. Apps and online resources can provide guidance for those new to these practices. Additionally, activities like listening to music, reading, or spending time in nature can also be relaxing and rejuvenating.

Balancing Work and Personal Life: This is essential for long-term stress management and resilience. This balance is about creating a boundary between work responsibilities and personal time. It involves managing workloads so that they don't encroach on personal time, which is needed for relaxation and rejuvenation. Flexible working arrangements, where possible, can help achieve this balance. It's also important to make time for personal interests, hobbies, and spending time with loved ones. Prioritizing self-care activities and ensuring regular 'unplugged' periods from technology can also contribute to a healthier work-life balance. Regularly assessing and adjusting this balance is necessary as personal and professional circumstances change.

Seeking Support and Resources: Recognizing when you need help and being open to seeking it is a sign of strength, not weakness. This support can come from various sources, such as professional counseling, employee assistance programs, or stress management workshops. Many organizations offer resources and support for stress management and mental health. Additionally, peer support groups, either within the workplace or externally, can provide a sense of community and shared understanding. Seeking guidance from mentors or colleagues who have experienced similar challenges can also be valuable. Asking for help is an important part of managing stress and maintaining emotional well-being.

Professional Development and Cross-Cultural Competence

Focused on broader professional development and cultural competence, this section covers advanced soft skills indispensable for IT experts. It probes into mastering public speaking and presentation skills, providing insights into effective communication and engagement with diverse audiences. Additionally, it addresses the importance of cultural competence in global environments, emphasizing the ability to collaborate and work effectively across cultural boundaries in today's interconnected world.

Public Speaking and Presentation Skills

The fundamentals of effective public speaking involve clear communication, tailored to your audience's level of understanding. Start by defining the purpose of your talk and understanding your audience's background and expectations. Clarity and conciseness are key; avoid jargon when speaking to non-technical audiences. Structuring your speech with a clear introduction, body, and conclusion can help maintain audience engagement. Practice is crucial; rehearse your speech to improve fluency and reduce nervousness. Engaging storytelling can make technical topics more relatable and memorable. Visual aids should complement your speech, not overshadow it. Finally, prepare for questions, as addressing them well can significantly enhance audience understanding and interaction.

Designing Compelling Presentations: Start with a clear outline of your main points, and design slides that support these points without overwhelming the audience with information. Use visuals like graphs, charts, and images to illustrate complex ideas. Consistency in fonts, colors, and layout enhances the professional look of your presentation. Storytelling can be a powerful tool; use narratives or case studies to make technical information more engaging. Keep text minimal on slides; they should act as cues for discussion rather than scripts. Practice presenting with your slides to ensure a seamless flow of information. The goal is to aid understanding, not to distract.

Overcoming Public Speaking Anxiety: This is common but can be managed with practice and preparation. Familiarize yourself with the venue and equipment beforehand to reduce anxiety. Breathing exercises and visualization techniques can help calm nerves before speaking. Focus on the message you want to convey, rather than on yourself. Remembering that the audience is generally supportive and interested in your content can also ease anxiety. Start with smaller groups and gradually build up to larger audiences to gain confidence. Seeking opportunities for public speaking, such as team meetings or small workshops, can provide valuable practice. Feedback from these sessions can guide improvements and build confidence.

Effective Use of Body Language and Voice: Maintain eye contact with the audience to create a connection and convey confidence. Use gestures to emphasize points, but ensure they are natural and not distracting. Your posture should project confidence; avoid crossing arms or slouching. Varying your voice's pitch, tone, and pace can keep the audience engaged and emphasize core points. Avoid speaking in a monotone; enthusiasm in your voice can make the topic more interesting. Practicing in front of a mirror or recording yourself can help you become aware of your body language and voice. Remember, non-verbal cues are as important as the words you speak.

Engaging the Audience: Start with a hook, such as a startling fact, a question, or a relevant anecdote to grab attention. Encourage audience participation through interactive elements like polls, questions, or discussions. Relating the topic to real-world applications can help the audience understand its relevance. Be attentive to the audience's body language and feedback; adjust your presentation style if needed. Pacing your speech and including short breaks or interactive segments can maintain interest. Summarize core points throughout to reinforce learning. Closing with a strong conclusion or call to action can leave a lasting impact.

Adapting to Different Audiences: Understand the audience's background: Are they technical experts, management, or non-technical stakeholders? Adjust the complexity of your language and the depth of technical details accordingly. For technical audiences, focus more on methodologies and technical insights. For non-technical audiences, emphasize the practical implications and outcomes. Be prepared to change your approach based on audience feedback and questions. Using relatable examples and analogies can bridge gaps in understanding. The ability to flexibly adapt your presentation ensures your message is effectively communicated, regardless of the audience.

Feedback and Continuous Improvement: Actively seek feedback from peers, mentors, or audience members after your presentations. Constructive criticism can provide insights into areas for improvement, such as clarity, pace, or engagement. Reflect on your performance: What worked well, and what could be better? Recording your presentations and reviewing them can offer a different perspective. Joining groups like Toastmasters or attending workshops can provide opportunities for practice and feed-back. Embrace a growth mindset; view each presentation as an opportunity to learn and improve.

Embracing Online Platforms for Public Speaking Mastery

In the digital age, technology offers innovative ways to refine public speaking skills from the comfort of your own home or office. Online platforms, harnessing the power of virtual reality (VR) and AI, provide an immersive experience that can mimic real-life speaking scenarios, offering personalized feedback and a wide range of practice environments. Here's how IT professionals can take advantage of these platforms:

VirtualSpeech is a VR application that combines realistic environments with AI feedback to help users practice public speaking, networking, and interview preparation. By using a VR headset, users can immerse themselves in various scenarios—from boardroom presentations to large conference halls. The AI analyzes performance, offering insights on pacing, use of fillers, eye contact, and engagement. This immediate feedback allows for rapid iteration and improvement. Benefits for individuals:

- **Tailored Feedback:** AI-generated insights help pinpoint specific areas for improvement, such as reducing filler words or improving eye contact, which are important for effective technical presentations.
- **Versatile Scenarios:** Users can practice in settings that mirror potential speaking engagements in their career, from team meetings to industry conferences.
- **Safe Practice Environment:** VR provides a risk-free space to practice and build confidence without the fear of real-world judgment.

Speeko is a mobile app designed to enhance public speaking skills through AI-powered feedback. The app offers exercises and tips to improve verbal communication, focusing on aspects like vocal clarity, pace, and confidence. Users receive real-time feedback on their speech delivery, helping them to identify and correct habits that may detract from their message's effectiveness. Benefits of Speeko:

- **Accessibility:** With the convenience of a mobile app, users can practice anytime, anywhere, fitting practice into their busy schedules.
- **Comprehensive Training:** Speeko covers a wide range of communication skills, from articulation to persuasive speaking, applicable in various IT contexts.
- **Personalized Progress Tracking:** The app tracks improvement over time, offering a tangible sense of progression and achievement.

By integrating platforms like VirtualSpeech and Speeko into their professional development toolkit, individuals can significantly enhance their public speaking skills. This not only boosts their ability to communicate complex ideas clearly and confidently but also positions them for greater success in leadership roles and client interactions. Incorporating these tools into daily or weekly routines can transform public speaking from a daunting task into a mastered skill, leveraging cutting-edge technology to accelerate personal and professional growth.

Cultural Competence in Global IT Environments

Understanding cultural diversity is vital in the global IT environment. It involves recognizing and respecting the different cultural backgrounds, beliefs, and practices of colleagues and clients worldwide. Awareness of cultural diversity leads to more effective communication and collaboration, as it fosters mutual respect and understanding. Being culturally aware can prevent misunderstandings and conflicts that arise from cultural differences. Training in cultural competence and exposure to diverse cultures can significantly enhance this understanding. Appreciating cultural diversity also means acknowledging the value that different perspectives and approaches bring to problem-solving and innovation in the industry.

Effective Cross-Cultural Communication: This requires an understanding of different communication styles, which can vary significantly across cultures. For instance, some cultures may value directness, while others prefer more indirect forms of communication. Being aware of and adapting to these differences is imperative. This also includes being mindful of language barriers and using clear, simple language when working with non-native English speakers. Using visual aids and confirming understanding through paraphrasing can also be effective. Sensitivity to non-verbal cues and respect for cultural norms in communication is crucial.

Cultural Sensitivity in Team Collaboration: This involves acknowledging and valuing the diverse backgrounds and viewpoints of team members. It's about creating an inclusive environment where everyone feels respected and their contributions are valued. This can involve flexible meeting times to accommodate different time zones or being mindful of cultural holidays and practices. Encouraging team members to share their cultural perspectives can enrich team dynamics and lead to more innovative solutions. It's important to address any cultural biases or stereotypes that may arise and foster a culture of open-mindedness and respect.

Global Mindset in Projects: This mindset involves thinking beyond local or national boundaries and understanding how global trends can impact projects. Embracing a global perspective can lead to more effective project management, particularly in terms of resource allocation, communication strategies, and risk management. Understanding the global implications of IT projects, such as data privacy laws across different countries, is also critical.

Continuous Learning and Cultural Awareness: Actively seeking opportunities to learn about different cultures, whether through formal training, travel, or personal research is advised. Engaging with colleagues from diverse backgrounds and being open to their perspectives can be a rich source of learning. Staying updated on global trends and how they impact different cultures can enhance this awareness. Continuous learning in cultural competence leads to more effective and respectful interactions in the global workplace.

In conclusion, this chapter underscores the importance of resilience, stress management, public speaking, and cultural competence in the professional growth of IT professionals. These skills are indispensable in navigating the challenges and complexities of the modern tech landscape. From building resilience against workplace stressors to mastering the art of public speaking and embracing cultural diversity, these skills enrich an IT expert's toolkit. They enhance personal effectiveness and foster healthier, more productive workplace relationships. As the IT sector continues to evolve, these soft skills will remain essential in adapting to new challenges and succeeding in a global, interconnected world.

Reflect and Apply Questions

1. **Resilience Building:** Reflect on a challenging situation in your life. How did you handle it, and what could you have done differently to show resilience?
2. **Stress Management Strategies:** Think about your current stress management techniques. Are they effective, and what new strategies could you implement to improve your well-being?
3. **Public Speaking Practice:** Imagine you have to explain a complex IT concept to a non-technical audience. How would you prepare, and what techniques would you use to ensure clarity and engagement?
4. **Cultural Sensitivity:** Recall an interaction with someone from a different cultural background. What did you learn, and how did it change your perspective or approach in a professional or personal context?

5. **Balancing Work and Life:** How do you currently balance your professional and personal life, and what strategies could you adopt to improve this balance?
6. **Learning from Feedback:** Consider a time when you received constructive feedback. How did you respond, and what did you learn from that experience?

Chapter 13: Ethical Implications and Responsible Use of Technology

In the evolving world of Information Technology, ethical computing has become a keystone of professional practice. This chapter explores the intricacies of ethical computing, an area that extends beyond mere compliance with laws and regulations. It involves a deep understanding of how our decisions and actions as IT specialists affect individuals, organizations, and society at large. We'll explore the principles that guide ethical decision-making, examine case studies of ethical dilemmas, and discuss the development of personal and organizational ethical standards. The goal is to equip you with the knowledge and tools to navigate complex ethical landscapes and uphold the highest standards of integrity and responsibility in your work.

Foundations of Ethical Computing

First, we will dig into the core principles guiding ethical behavior in the digital realm, exploring concepts like integrity, accountability, and transparency. This section lays the ground-work for understanding the ethical responsibilities inherent in IT professions, providing insights into navigating complex ethical dilemmas and upholding professional standards.

Understanding Ethical Computing

Ethical computing revolves around a set of principles and values that guide IT specialists in their decision-making processes. These principles include honesty, fairness, respect for user privacy, and responsibility for the consequences of technological actions. Understanding these core values is important for making ethical choices that align with both personal and professional standards. Ethical computing is not just about doing what is legally right but also what is morally responsible. These values often act as a compass in situations where regulations may not provide clear guidance, ensuring that IT practices contribute positively to society.

Case Studies and Historical Perspectives: The Cambridge Analytica scandal is a watershed moment in the discussion of digital ethics, data privacy, and the role of technology in democratic processes. In 2018, it was revealed that Cambridge Analytica, a political consulting firm, had accessed the personal data of millions of Facebook users without their consent. This data was used to influence voter behavior in the 2016 U.S. presidential election and the Brexit referendum. The scandal raised serious questions about the responsibility of tech companies in safe-guarding user data, the importance of consent in data collection, and the impact of data misuse on democracy. It stressed the need for stringent data protection policies, transparency in data usage, and the ethical obligation of tech companies to prevent the exploitation of their platforms.

The development of AI and autonomous vehicles presents complex ethical dilemmas, particularly in the programming of ethical decision-making into machines. Autonomous vehicles, for instance, must make split-second decisions in critical situations that could potentially harm passengers, pedestrians, or other drivers. The challenge lies in creating algorithms that can navigate these ethical dilemmas, such as deciding between the lesser of two harms in an unavoidable accident. This presents the necessary ethical considerations in AI development, including the potential for AI bias, the importance of transparent decision-making processes, and the need for a multidisciplinary approach to ethical AI development. There is an ongoing debate among technologists, ethicists, and policymakers on establishing ethical guidelines for AI and autonomous systems.

Exploring case studies and historical perspectives enriches our understanding of ethical computing's complexities. These narratives shed light on the repercussions of unethical practices and illustrate the evolving nature of ethical guidelines in response to technological advancements and societal shifts. From the significant impact of data privacy violations to the ethical quandaries posed by AI and autonomous systems, each case study serves as a learning opportunity. They underscore the importance of ethical vigilance in technology development and deployment, offering valuable lessons to individuals committed to upholding ethical standards in their work.

Ethical Decision-Making Models: These models typically involve steps like identifying the ethical issues, considering the stakeholders affected, evaluating the options from a moral stand-point, and making a decision that aligns with ethical principles. Familiarity with these models aids individuals in systematically navigating complex ethical situations. They provide a framework for considering the broader implications of technological decisions and help in achieving a balance between tech advancement and ethical responsibility.

Impact of Unethical Computing Practices: Unethical practices can lead to serious consequences, including legal repercussions, loss of public trust, and harm to individuals and society. Issues like data manipulation, privacy breaches, and intellectual property theft not only damage organizational reputations but also have broader societal implications. This awareness stresses the responsibility of IT professionals to uphold ethical standards in their work. Recognizing the potential negative outcomes of unethical behavior motivates a commitment to ethical computing practices.

Developing Personal Ethical Standards: Developing these standards involves reflecting on one's values and beliefs and how they align with ethical practices in technology. Personal ethical standards act as a guide in making decisions, especially in situations where external guidelines may be ambiguous or lacking. Continuous self-reflection and learning about ethical issues in IT help in refining these standards over time. It's also important to be open to diverse perspectives and ethical viewpoints, as this openness can enrich one's understanding of complex ethical landscapes in the industry.

Ethics in Software Development: Ethical considerations in software development include respecting user privacy, ensuring data security, and avoiding harm through software functionality. Ethical software development also involves being transparent about data usage, obtaining user consent, and considering the societal impact of software products. As technology becomes more integrated into daily life, the ethical implications of software decisions become more significant. Software developers must therefore integrate ethical considerations into every stage of the development process, from design to deployment.

Cybersecurity Ethics

Ethical hacking and penetration testing are practices in cybersecurity, conducted with the intention of identifying and fixing security vulnerabilities. Unlike malicious hacking, ethical hacking is authorized and aims to improve system security. Ethical hackers adhere to legal guidelines and respect privacy and data integrity while testing. Clear agreements regarding the scope and methods of testing are necessary to maintain ethical standards. Transparency in reporting findings and recommending improvements is also an important aspect of ethical hacking. Taking this proactive approach helps strengthen cybersecurity defenses.

Data Security and Individual Rights: Protecting user data is not just a technical issue but also an ethical obligation. It involves ensuring that personal information is secured against unauthorized access and breaches. Respecting individual rights means considering the privacy and confidentiality of user data in all cybersecurity measures. Ethical challenges arise in balancing the need for robust security with respecting users' privacy and autonomy. Developing cybersecurity solutions that uphold individual rights requires a commitment to ethical practices and user-centric approaches.

Corporate Responsibility in Data Protection: In cybersecurity ethics, companies must responsibly handle data protection. It's their duty, both legally and morally, to safeguard user data. This involves putting robust security practices in place, consistently updating systems, and educating staff on protecting data. Ethical business conduct also requires clear user communication about data use and swift action and disclosure in case of data breaches. Such commitment is key to preserving user trust and the organization's reputation.

Ethical Dilemmas in Cybersecurity: Cybersecurity presents various ethical dilemmas, such as the balance between security measures and user privacy. Decisions about data access, surveillance, and information sharing often involve complex ethical considerations. Addressing these dilemmas requires a careful evaluation of the potential impacts on stakeholders and society at large. Ethical decision-making in cybersecurity calls for a deep understanding of both technological aspects and moral principles. Developing ethical guidelines and frameworks can aid cybersecurity professionals in navigating these challenges.

Cybersecurity Policies and Governance: These policies provide a framework for managing and securing information assets, ensuring compliance with legal and ethical requirements. Effective governance involves leadership commitment, clear communication of policies, and regular monitoring and review. It also includes establishing accountability for cybersecurity within the organization. Ethical governance ensures that cybersecurity practices are not only effective but also align with broader ethical and societal values.

Ethical Implications of Surveillance: While surveillance technologies can enhance security, they also raise concerns about privacy invasion and potential misuse. Ethical surveillance practices involve ensuring that monitoring is justified, proportionate, and transparent. It's important to consider the impact of surveillance on individuals' rights and freedoms. Challenges arise in determining the extent to which surveillance is necessary for security and where to draw the line to protect individual privacy. In responsible surveillance practices, it's imperative to find the right balance between security requirements and ethical concerns.

Professional Responsibility in IT

Upholding professional ethics is fundamental to maintaining trust and credibility in the field. This involves adhering to moral principles like honesty, fairness, and integrity in all professional endeavors. IT experts are expected to make decisions that not only benefit their organization but also consider the welfare of users and society. Ethical behavior includes avoiding conflicts of interest, ensuring accuracy and truthfulness in work, and respecting intellectual property rights. Commitment to these practices builds a foundation of trust with colleagues, clients, and the public.

Confidentiality and Trust: Individuals are often privy to sensitive information, and it's imperative to safeguard this data diligently. This responsibility involves implementing robust security measures and adhering to privacy laws and company policies. Maintaining confidentiality also means being discreet about internal matters and respecting the privacy of clients and users. Breaches of confidentiality can lead to significant legal and reputational consequences and erode trust in both the individual and the organization.

Accountability in IT Practices: This entails taking responsibility for one's actions and decisions in the professional realm. IT experts must be accountable for ensuring their work meets the required standards and does not harm users or society. This includes being transparent about mistakes or failures and taking steps to rectify them. Accountability also means staying informed about the latest industry standards and practices and adhering to them. A culture of accountability within organizations promotes trust, reliability, and ethical behavior.

Ethical Dilemmas and Decision Making: Ethical dilemmas often involve complex situations where the right course of action is not clear-cut. Decision-making in such scenarios requires careful consideration of ethical principles, the potential impact on stakeholders, and legal implications. Individuals may face dilemmas related to user privacy, data management, or emerging technologies. Developing a systematic approach to ethical decision-making, such as consulting industry guidelines, seeking advice from peers, or following ethical frameworks, can help navigate these challenges.

Professional Codes of Conduct: These codes provide guidelines for ethical and responsible behavior in the profession. They often cover aspects like professional competence, integrity, respect for privacy, and avoiding harm. Adherence to these codes is important for maintaining professional standards and public trust. IT organizations and associations typically establish these codes, and they serve as a reference point for ethical decision-making and professional behavior.

Legal Responsibilities: This includes data protection laws, intellectual property rights, and contractual obligations. Individuals must ensure that their work adheres to these legal requirements, which can vary across different jurisdictions. Staying informed about relevant laws and seeking legal advice when necessary is part of fulfilling these responsibilities. Ethical practice involves legal compliance and considering the broader ethical implications of one's work.

Ethical Leadership in Teams: Ethical leadership involves guiding and inspiring team members to uphold high ethical standards in their work. These leaders model integrity and transparency, setting the tone for ethical behavior within the team. They encourage open communication, foster a culture of accountability, and provide guidance on ethical issues. Ethical leaders also prioritize the development and well-being of their team members, promoting a positive and inclusive work environment. Their leadership style contributes to building a team that values and practices ethical and responsible IT.

Regulatory Compliance and Ethics

To ensure that IT practices adhere to legal and ethical norms it is imperative to understand compliance standards. These standards, like ISO/IEC 27001 for information security management, provide guidelines to ensure IT operations are secure and reliable. These standards cover various aspects, including data protection, cybersecurity, and risk management. Adhering to them ensures legal compliance and helps in maintaining ethical standards. Individuals must be familiar with these standards, understand their requirements, and implement them effectively in their operations.

Legal Obligations: These involve adhering to laws and regulations that govern technology use and data management. This includes compliance with data protection laws like GDPR, intellectual property laws, and industry-specific regulations. Legal compliance is paramount to avoid penalties and maintain organizational integrity. It also involves contractual obligations with customers and partners. IT experts have a responsibility to stay updated with current and future legal requirements and incorporate them into their practices.

Ethical vs. Legal Obligations: Navigating the intersection of ethical and legal obligations is a complex task. While legal compliance is mandatory, ethical obligations often extend beyond legal requirements, focusing on doing what is morally right. For instance, a company may legally collect user data, but ethically, it should consider the privacy implications of how it uses that data. Ethical obligations include transparency, fairness, and responsibility towards stakeholders. Balancing ethical and legal obligations involves making decisions that not only comply with the law but also uphold high moral standards.

Compliance in Cloud Computing: Cloud providers and users share the responsibility for compliance, known as the shared responsibility model. Users must understand the scope of their responsibility, which varies depending on the cloud service model (IaaS, PaaS, SaaS). Regular audits and assessments are necessary to ensure ongoing compliance in the cloud environment.

Impact of Compliance on IT Strategy: Regulatory requirements can influence priorities and investments. Compliance considerations may drive decisions around data management, security infrastructure, and software development. An effective IT strategy should incorporate compliance as a core component, aligning it with business objectives and risk management. This involves proactive planning to accommodate current compliance needs and the flexibility to adapt to changing regulations.

Social and Environmental Considerations

In this section, we will examine the broader implications of technology on society and the environment. From addressing issues of privacy and data protection to promoting sustainable computing practices, this section highlights the connection between technological advancements and societal well-being. By exploring ethical considerations in emerging technologies and regulatory compliance, you will gain a comprehensive understanding of the responsible use of technology in today's world.

Privacy and Data Protection

Data privacy principles form the backbone of ethical data management. These principles include the fair and transparent collection of data, ensuring accuracy, and upholding individuals' rights to access their data. Respecting data minimization, where only necessary data is collected, is also crucial. Data privacy involves protecting personal information from unauthorized access and misuse, ensuring that individuals maintain control over their own data. Understanding and implementing these principles is vital for IT professionals to navigate the complex landscape of data privacy, maintaining user trust and compliance with regulatory standards.

Implementing Data Protection Measures: Effective data protection measures are critical in safeguarding sensitive information in IT environments. This involves deploying robust security technologies such as firewalls, antivirus software, and intrusion detection systems. Regularly updating and patching software and systems is necessary to protect against vulnerabilities. Employee training and awareness programs are also recommended to prevent data breaches caused by human error or insider threats. Regular audits and risk assessments can help identify potential security gaps. Data encryption, secure data storage, and controlled access are additional practices in ensuring data security.

Impact of Data Breaches: Breaches often lead to exposure of sensitive personal information, which can result in identity theft and fraud. They can also have regulatory implications, leading to hefty fines and legal action. Beyond the immediate effects, data breaches can erode customer confidence and trust in an organization. They highlight the need for robust data protection measures and proactive risk management strategies. Understanding the potential impacts can motivate IT experts to prioritize data security in their practices.

Ethical Data Collection Practices: This ensures that data is gathered responsibly, respecting individuals' privacy and rights. It involves obtaining explicit consent from individuals before collecting their data, being transparent about the purpose of data collection, and ensuring the data is used only for its intended purpose. Ethical practices also include anonymizing data where possible and ensuring data accuracy. It's important to avoid collecting unnecessary data, adhering to the principle of data minimization. Ethical data collection is a legal requirement and a moral obligation to protect individuals' privacy.

User Consent and Transparency: Consent must be freely given, specific, informed, and unambiguous. This involves clearly informing users about what data is being collected, for what purpose, and how it will be used. Opt-in mechanisms should be clear and easily accessible. Transparency also extends to data sharing practices and data retention policies. Regularly updating privacy policies and making them easily accessible contributes to transparency. IT professionals must ensure that users have control over their data, including the ability to access, correct, or delete it.

Navigating Global Privacy Laws: Different countries and regions have varied regulations, such as GDPR in Europe and CCPA in California. IT specialists should understand these laws especially in organizations that operate internationally. Compliance involves adhering to legal standards and aligning with ethical practices. This might include implementing stricter data protection measures than required by law to ensure user privacy. Staying informed about evolving privacy laws and adapting practices accordingly is necessary to avoid legal penalties and uphold ethical standards.

Role of Encryption in Data Protection: Encryption serves as a fundamental tool for securing sensitive information. It involves converting data into a coded format that can only be accessed with the correct encryption key. Encryption protects data both in transit and at rest, making it mandatory for securing online communications, stored data, and cloud services. Implementing strong encryption standards is a best practice in safeguarding against unauthorized access and data breaches. IT pros must understand various encryption methods and apply them appropriately to protect sensitive information.

Privacy in Cloud Computing: Ensuring data privacy in the cloud involves choosing reputable cloud service providers with robust security measures. This includes data encryption, secure data centers, and regular security audits. Understanding the shared responsibility model in cloud security, where both the provider and the user have roles in protecting data, is important. Compliance with data protection laws and ensuring data is stored and processed in accordance with privacy regulations are core considerations. Transparent data governance policies and practices also play a big role in maintaining privacy in cloud computing environments.

Technology and Society

Technological advancements have profoundly impacted social change, reshaping how we communicate, work, and live. Innovations like the internet, mobile technology, and social media have transformed human interaction, creating a more connected world. While these advancements offer numerous benefits, such as increased access to information and ease of communication, they also present challenges, including privacy concerns and the potential for misinformation. The rapid pace of technological change requires careful consideration of its social implications, ensuring that progress benefits society as a whole. Addressing the ethical aspects when launching new technologies is key to safeguarding society from potential negative consequences.

Digital Citizenship and Online Behavior: Digital citizenship encompasses the norms of appropriate, responsible behavior concerning technology use. It's about understanding the rights and responsibilities that come with digital participation. Good digital citizenship involves respecting others online, understanding the permanence of digital footprints, and using technology in a way that positively contributes to society. Issues like cyberbullying, online privacy, and digital ethics are important to this concept. Promoting digital citizenship is imperative to creating a safe and respectful online environment.

Tech's Role in Enhancing Accessibility: Advances in assistive technologies, such as screen readers, voice recognition software, and customizable interfaces, have made technology more accessible. However, there's a continuous need for innovation and improvement in this area. Ethical considerations in technology design must include a focus on accessibility, ensuring that all users have equal access to digital resources. Individuals are challenged to create solutions that are not only technologically advanced but also universally accessible.

Technology's Impact on Employment: This offers both opportunities and challenges. While automation and AI have led to job displacement in certain sectors, they have also created new job opportunities in others. The ethical challenge lies in addressing the skills gap and ensuring that workers are trained for the jobs of the future. It's also important to consider the implications of remote work technologies on employment patterns and workplace cultures.

The Digital Divide and Global Access: This refers to the gap between those who have easy access to the internet and technology and those who do not. This divide can be due to various factors, including socioeconomic status, geography, age, and education. Bridging this divide is vital for ensuring equitable access to technology, which is increasingly necessary for participation in modern society. The ethical imperative is to work towards global digital inclusion, ensuring that the benefits of technology are available to all.

AI and Ethical Considerations

AI bias is an ethical issue where AI systems may exhibit biased behaviors or decisions, often reflecting the biases present in their training data. This bias can lead to unfair or discriminatory outcomes, particularly in sensitive areas like hiring, law enforcement, and lending. Addressing this bias involves understanding its sources, such as biased data sets or flawed algorithmic design. Ethical AI development requires proactive measures to identify and mitigate bias, including diverse data sets and testing algorithms for fairness. Continuous monitoring and evaluation are required to ensure AI systems operate without perpetuating existing biases.

Ethical AI design and development encompass creating AI systems that adhere to moral principles and values. This involves considering the potential impacts of AI on individuals and society from the outset. Ethical AI development includes ensuring transparency, accountability, and fairness in AI systems. It also requires involving diverse stakeholders in the design process to understand different perspectives and potential impacts. Ethical guidelines and frameworks can guide developers in creating responsible AI technologies. Regular ethical evaluations and audits help uphold the integrity of these systems.

AI in Surveillance and Privacy: AI-enabled surveillance systems, like facial recognition technologies, have the potential for misuse or abuse, leading to privacy violations. Balancing the benefits of AI surveillance for security with the need to protect individual privacy rights is a major ethical challenge. Implementing strict regulations and oversight, ensuring transparency in usage, and allowing opt-out options are core steps. Ethical AI surveillance should respect individual privacy while serving legitimate and lawful purposes.

Future Challenges: The future of AI ethics involves anticipating and addressing the evolving challenges as AI technologies advance. This includes dealing with advanced AI capabilities like deepfakes or AI-generated content, which have implications for truth and trust in digital media. Another challenge is ensuring equitable access to these technologies, preventing the widening of existing societal divides. The potential for AI to influence human behavior and decision-making also raises ethical questions about autonomy and manipulation. Preparing for these future challenges requires ongoing dialogue, research, and collaboration among technologists, ethicists, policymakers, and other stakeholders.

Sustainable Computing

Sustainable IT involves adopting practices that minimize the environmental impact of computing technologies. This includes designing, manufacturing, operating, and disposing of computers and computer-related products in environmentally friendly ways. Core principles include energy efficiency, reducing resource consumption, and minimizing e-waste. Sustainable IT also involves considering the entire lifecycle of technology products, from production to disposal. Emphasizing renewable energy sources and environmentally friendly materials is part of this principle. Embracing these principles helps minimize the environmental impact of IT operations and supports a more sustainable future.

Eco-Friendly Hardware Choices: This involves selecting hardware that is energy-efficient, durable, and easy to recycle. Manufacturers are increasingly focusing on creating environmentally friendly products, including using recycled materials and designing for easy disassembly. Consumers and IT specialists can contribute by choosing devices with a lower environmental impact and considering the sustainability practices of manufacturers. Proper maintenance and extended use of hardware also reduce environmental impact.

Sustainable Software Development: This process focuses on creating software that is efficient, durable, and has a minimal environmental impact. This involves optimizing code for energy efficiency, particularly in applications that run on servers or data centers consuming significant power. Sustainable software practices also include considering the software's lifecycle, designing for longevity, and avoiding planned obsolescence. Encouraging open-source contributions can lead to more sustainable software solutions by fostering community involvement and resource sharing.

Energy Efficiency in Data Centers: Data centers consume a significant amount of energy, making energy optimization crucial. This can be achieved through advanced cooling systems, energy-efficient server design, and optimizing data center layouts. Using renewable energy sources for data centers can also significantly reduce their environmental impact. Implementing practices such as server virtualization and cloud computing can lead to more efficient use of resources. Regularly monitoring and auditing energy use is essential for identifying opportunities for further efficiency improvements.

Recycling and Disposal of Tech Products: E-waste contains hazardous materials, which, if not properly disposed of, can harm the environment and human health. Recycling programs help recover valuable materials and prevent harmful substances from entering the environment. IT professionals and consumers should be aware of and utilize proper e-waste disposal channels. Advocating for and adopting policies that encourage recycling and responsible disposal within organizations is also recommended. Additionally, supporting companies that take back and recycle their products at the end of their lifecycle can drive industry-wide change.

Promoting Green IT Practices: This can be achieved through training programs, sustainability initiatives, and advocacy within organizations. Encouraging the use of energy-efficient devices, promoting remote work, and advocating for sustainable software development are ways to foster green practices. Organizations can also implement policies that prioritize sustainability in IT procurement and operations. Collaboration with environmental experts and participation in sustainability forums can further enhance the adoption of green practices. Promoting these practices contributes to a more sustainable and environmentally responsible industry.

Digital Accessibility

The legal aspects of digital accessibility are governed by various laws and regulations, which aim to ensure that digital products and services are accessible to all, including people with disabilities. Key legislations include the Americans with Disabilities Act (ADA) in the U.S. and the Equality Act in the UK. These laws mandate that digital content, especially on public platforms, must be accessible. Non-compliance can lead to legal challenges and penalties. It is recommended for IT specialists to understand these legal requirements to ensure their digital products meet accessibility standards.

Tools for Creating Accessible Content: These tools help identify and fix accessibility issues in web content and applications. They range from accessibility checkers and validators that test for compliance with standards like WCAG, to software that simulates how people with different disabilities interact with digital content. Employing these tools in the development and testing phases is important for ensuring that digital products are accessible to all users. They provide guidance and automate some aspects of accessibility testing, making it easier for developers to create inclusive content.

Inclusivity in User Interface (UI) Design: This involves considering various user needs and preferences in the design process. Inclusive UI design includes using clear language, providing alternate ways to consume and interact with content, and ensuring that color and contrast are used effectively for users with visual impairments. Designing with inclusivity in mind not only meets ethical and legal standards but also reaches a wider audience, enhancing the overall user experience.

Best Practices for Digital Accessibility: Best practices include following established guidelines like WCAG and ensuring continuous evaluation and improvement of digital products. It's advised to involve people with disabilities in the design and testing process. Accessibility should be a consideration from the start of the design process, not an afterthought. It is encouraged to ensure regular training for designers and developers in accessibility standards and practices. Documentation and clear communication of accessibility features also help users utilize these features effectively. Collaborating with accessibility experts can provide additional insights and guidance.

Ethics in Emerging Technologies

Blockchain technology, known for powering cryptocurrencies, presents unique ethical implications. Its decentralized nature raises questions about accountability and governance. The transparency of blockchain can conflict with privacy needs, as seen in public ledgers. Ethical use of blockchain requires balancing its potential for transparency and security with privacy and data protection concerns. Additionally, the environmental impact of blockchain, particularly energy-intensive processes like Bitcoin mining, is a growing concern. Addressing these issues involves developing sustainable, privacy-respecting blockchain solutions.

IoT Security and Privacy Issues: The Internet of Things (IoT) has transformed everyday objects into interconnected smart devices, raising significant security and privacy issues. With IoT devices collecting vast amounts of data, there is an increased risk of privacy breaches and unauthorized data access. Ensuring robust security measures and protecting user data is a major ethical responsibility. This involves designing IoT devices with security in mind, regularly updating software, and being transparent with users about data collection practices.

Ethical Use of Drones and Robotics: These technologies are advancing rapidly, presenting various ethical challenges. Issues include privacy concerns, as drones can collect data in public and private spaces. There's also the ethical debate around the use of drones and robotic technology in military and surveillance applications. Ensuring these technologies are used in ways that respect privacy and ethical norms is important. This includes developing regulations and guidelines for drone usage and considering the societal impact of robotics, particularly in terms of employment and human interaction.

Quantum Computing and Data Security: Quantum computing's potential to break traditional encryption methods raises momentous ethical concerns regarding data protection. Preparing for a post-quantum cryptography era is critical to protect sensitive information. This involves researching and developing quantum-resistant encryption methods. Ethical considerations include balancing the advanced capabilities of quantum computing with the need to protect data against future quantum attacks, ensuring long-term data security and privacy.

Ethical Challenges in Virtual Reality: Challenges include the potential for causing psychological or emotional harm, privacy issues related to data collection in VR environments, and the potential for creating addictive or disorienting experiences. Ethically developing and deploying VR technology involves creating content responsibly, ensuring user safety and privacy, and being mindful of the psychological impacts. Guidelines for ethical VR content creation and use are needed to navigate these challenges.

Biometrics and Privacy Concerns: While biometrics offer enhanced security, they also involve collecting and storing sensitive personal data. The ethical handling of biometric data involves ensuring robust security measures to prevent data breaches and misuse. It's important for users to know how their biometric data is handled and stored. Balancing the convenience and security benefits of biometrics with the need to protect individual privacy is a huge ethical challenge in this field.

Ethical Frameworks for New Technologies: These frameworks provide guidelines and principles for addressing ethical challenges posed by emerging technologies. They involve considering the societal, environmental, and individual impacts of technology. Including diverse perspectives in the development of these frameworks ensures a comprehensive approach to ethical considerations. Regularly revising and updating ethical frameworks is necessary to keep pace with technological advancements, ensuring that ethical guidelines remain relevant and effective.

In conclusion, Chapter 13 has explored the vast and complex terrain of ethics in IT. From the challenges posed by emerging technologies like AI and blockchain to the importance of data privacy and cybersecurity, we've examined how ethical considerations are integral to the profession. This chapter emphasized the balance between innovation and ethical responsibility, stressing the importance of thoughtful decision-making in technology. As the digital world continues to evolve, IT professionals must remain vigilant, informed, and committed to upholding the highest standards of ethical practice.

Reflect and Apply Questions

1. **Ethical Decision-Making:** Reflect on a situation where you had to make a difficult decision. How would you apply ethical principles to make a similar decision in the future?
2. **Privacy and Data Protection:** Consider your online behavior. How might understanding data privacy laws influence how you manage your personal data or approach projects in IT?
3. **Balancing Ethics with Innovation:** Think of an innovative technology you've recently read about. What ethical considerations do you think should be addressed in its development or deployment?
4. **Cybersecurity Awareness:** Reflect on your understanding of cybersecurity. How can awareness of ethical

issues in cybersecurity influence your actions as an IT professional?

5. **Digital Accessibility:** How can you contribute to making digital content more accessible, based on what you've learned about digital accessibility and its importance?

Conclusion

Embracing the IT Journey

The initial stages of the book introduced the expansive world of IT, laying out a broad yet nuanced landscape of this dynamic field. These chapters helped to clarify Information Technology, making it accessible to newcomers. They focused on the vast array of opportunities in the industry, guiding readers through the various pathways they can explore. Emphasis is placed on understanding the importance of foundational concepts and their real-world applications. This introduction serves as a springboard, encouraging readers to dive deeper into the specific areas that pique their interest, and setting the stage for a rewarding journey into the realm of technology.

Navigating Advanced Technologies

As we continued, the book transitions into advanced topics, signifying the constant evolution of technology and the need for IT professionals to adapt and grow. Covering areas like cybersecurity and cloud computing, these chapters offered both depth and breadth, revealing the intricacies and potentials of modern technologies. They served as a gateway to specialized fields, urging readers to explore beyond the basics and embrace the complexities and challenges of the tech world.

Preparing for a Sustainable Future

There are sections dedicated to looking forward, contemplating the future trajectory of IT careers. They discuss the importance of agility and lifelong learning in an ever-changing industry, emphasizing that staying relevant requires continuous adaptation and growth. These sections inspire readers to be forward-thinking, not just in terms of technology, but also in considering the broader impact of their work on society and the environment. This forward-looking perspective is key for cultivating a sustainable, responsible, and impactful IT career.

The Human Element in IT

In the realm of Information Technology, technical skills are just one part of the equation. The book dedicates significant atten-tion to soft skills — the human element that complements technical expertise. This focus acknowledges that effective communication, teamwork, and adaptability are pivotal in shaping a well-rounded IT specialist. These chapters reinforce the idea that success is as much about interpersonal skills and emotional intelligence as it is about technical knowledge.

Ethical and Social Responsibilities

The later chapters bring to light the ethical dimensions and societal implications of working in IT. These sections are critical, as they address the responsibility of individuals to make decisions that are not only technologically sound but also morally and ethically grounded. The discussions on ethics and social responsibility highlight the influential role of IT in shaping societal norms and values, emphasizing the need for a conscientious approach to technology.

Call to Action for Growth and Innovation

As you close the pages of this book, remember that a journey in Information Technology is not just a career path, but a journey of continuous learning and growth. This dynamic world awaits your unique contributions and innovations. Don't hesitate to take the first step, whether it's enrolling in a course, experimenting with a new technology, or simply networking with professionals in the field. Embrace the challenges and opportunities the industry offers. Let your curiosity drive you, and let your passion be the guiding force in your pursuit of technological mastery. The future is not just about technology; it's about the people who shape it. Be bold, be creative, and most importantly, be ready to transform the world with your skills and insights. Your journey in IT starts now – seize the moment and make it count!

Want more?

Check out my website www.unitethegrid.com. We are the GRID, a collective of IT professionals that aim to provide all of the resources you need to excel in your IT career in one location. We offer books, study sheets, mentoring, coaching, recommendations, and more to come! Come check us out and accelerate your path to a rewarding and successful career in IT! If you found this book helpful along your journey, I encourage you to return to the platform where you obtained the book to leave a good review. This will help others find the book, allowing it to help a wider audience.

Appendix A: Acknowledgements

As I sit down to write this acknowledgement, I am overwhelmed with a sense of gratitude. This book is not just a product of my efforts but a testament to the support and encouragement I've received from so many. It is with a full heart that I extend my deepest thanks to those who have been part of this journey.

First, I want to thank my wife, Meghan, for always being there for support and recommendations when I need it. I would not be where I am today without you in my life. I love you and appreciate all that you do. I am grateful for having you by my side along this journey.

To my kids, Emma and Logan, you both are my world. You have given my life purpose from the first time I held each of you. Whether you know it or not, you are the reason I wake up everyday and strive to be a better father and person. I love you both dearly!

To my parents and siblings, there is so much to say here. My father is the reason I got into technology. I remember when I was very young, he would build computers and I was fascinated at how they worked together. My mother played a huge role in raising me being a predominantly stay at home mom. My brothers and sisters, I have many cherished memories with each of you, and I value our relationships. I truly hope I impact your lives half as much as you have impacted mine over the years. I couldn't ask for a better family of support. I love you all!

To my grandmother, late grandfather, and uncle, you have been supportive of me for my entire life. I am blessed to have so many wonderful memories with you. I love you all and truly thank you for everything you have done for me.

Professionally, I have to thank my mentors thus far, as I would not be where I am without their guidance. Matt, you are my newest mentor and we have built a solid relationship. Having you in my corner has been paramount for my career advancing as it has.

To Jason, I will never forget the subtle kick you gave me to get started with my CCNA. Ever since that day, I have been on a relentless pursuit to be the best person I can be and help others along the way. I truly believe our relationship was a turning point in my career.

There are many others who had an impact on my life and journey thus far, too many to name. To each of you that has helped me through guidance, support, or even casual conversations, I thank you for being there. Now it is my turn to give back and help others the way you have helped shape me.

To my beta readers, each of you were important to this book. Your insights, suggestions, and critiques helped me fine tune this to make a better, stronger book to help others. Thank you all!

Last but not least, I have to thank God for giving me this opportunity. I have been blessed beyond belief across my life with a supportive family, caring friends, helpful mentors, and great colleagues. I intend to use my platform to give back to others struggling to find their way into the industry or unhappy with their career path. This is the beginning of something great!

Appendix B: Recommendations

Disclaimer: The recommendations for training sites, apps, and other resources provided in this section are based solely on my personal opinion and experience as of the time of writing this book. These recommendations are not endorsements and I do not receive any compensation for mentioning them. The effectiveness and relevance of these resources may vary, and I encourage readers to exercise their own judgment and conduct due diligence before using any of these suggested resources. Please be aware that the field of information technology is rapidly evolving, and newer resources may become available after the publication of this book.

Throughout my journey, I have spent countless hours studying, researching, and testing out different resources. Many of those would have been beneficial early on in my career. There are still so many tools out there that I haven't used yet. This is a list, in no particular order, of resources in different categories that I would recommend you research and test out.

Learning Platforms: Free and Paid

- **YouTube** – One of the most underrated tools in my opinion. There are so many professionals out there who are not as well-known but publish content. In my experience, it can be beneficial to hear a concept I am having trouble grasping explained differently. You can even find short videos digging into specific topics. For instance, I struggled differentiating routing protocols. I went to YouTube and found many videos on specific routing protocols, and comparing/contrasting ones I was struggling with. This assisted my learning and helped me pass one of my exams
- **LinkedIn Learning** – This resource has a vast catalog of training material. Not just for IT related topics either. There is a ton of content for helping improve soft skills which can be invaluable. This is a great investment that comes with a broad amount of content to learn from.
- **Udemy** – This is probably the learning resource I have used second most to YouTube. This site has frequent deals where you can get courses at a great price. What I love about Udemy is they have many courses that include labs or additional resources. I took an AWS cloud course that utilized a free

AWS account to use with the lectures. This allowed for hands on practice of the techniques taught in the course. That helped to retain the information.

- **CYBRARY** – CYBRARY has both a free and paid plan. Their free plan includes a large collection of relevant IT learning material. It is a good idea to take advantage of all of the resources available to you, especially free ones. Check out their library for yourself and see what you think.
- **Hack The Box and Try Hack Me** – These sites are known for their hands on labs built to train ethical hackers. When I first experienced each, I was amazed at how much more they really had. There are Cybersecurity labs, SIEM labs, and much more. It is a gaming type of environment as well. If you aren't a book person, either of these can be great for learning basics and advanced topics.

Hardware/Gadgets

- **Raspberry Pi** – I didn't get my hands on one of these until well into my career and I wish I had of much earlier. A Raspberry Pi is a small piece of hardware components capable of running operating systems like Linux. There are so many cool projects that can be done with a Pi like creating an adblocker, creating a proxy, even using it as a personal computer. This is an excellent learning tool and there is a huge world of videos and tutorials available. If you like getting hands on and learning, this may be the gadget for you.
- **Hacking Tools** – If ethical hacking is of interest to you, there is an arsenal of tools you can get your hands on to practice with. Tools like a Rubber Ducky, Flipper Zero, Wi-Fi Pineapple, and even cables that can be used for ethical hacking. Once again, this is another tool set to get hands on and practice with, ethically of course!
- **Virtual Labs** – Software is available that can be ran on your computer giving you the ability to create virtual desktops. This can be excellent for learning a new OS like Linux, or even creating a cost-effective virtual lab without the physical equipment. One I have used is VirtualBox. If you plan to work on the networking side of IT, a home lab can be extremely beneficial to configure and troubleshoot when things go wrong.

Productivity Apps

- **Notes** – As suggested many times throughout this book, getting into the practice of diligent note taking is valuable. While there are many apps available, the one I use at the time of writing this book is OneNote. It allows you to create different notebooks, chapters, pages, and more. It also translates well from a desktop to mobile device. It is convenient for me as I can separate exam notes, review, and anything else relevant. I highly suggest doing research and testing out some note taking apps.
- **Scheduling** – Keeping a schedule can help you plan your day ahead and keep you on track. Once again, there are many options out there to choose from. Personally, I use Microsoft Outlook. It keeps my email integrated with my schedule, and I can even integrate a to do list. It definitely helps being able to look at my week ahead and see when I will be busy and when I have time blocks for additional learning, personal appointments, or free time activities. With an array of different tools, do your research and find one that works best for you!
- **Mind Mapping Tools:** Apps like MindMeister for brainstorming and organizing thoughts can be useful for complex IT concepts.

Communication & Collaboration Tools

- **Zoom/Teams** – These are 2 of the more common meeting apps. Either of these will be beneficial in keeping connected when working with a global team, when working remotely, and for virtual meetings or interviews. If you haven't already, I recommend looking into one of these or a similar app as they are becoming more widely used due to the growing nature of remote work.
- **Slack/Teams** – Both of these are great for instant messaging, sharing images or links, etc. They are also great for helping keep connected with team members in different parts of the country or world. One feature I love about both is the ability to integrate other technologies into them. For instance, RSS feeds. They can be integrated so you have a Teams or Slack channel that provides relevant industry news or updates.

- **GitHub** – This is an excellent platform to collaborating in software development. It's also a valuable resource for IT professionals to explore existing projects, learn from other developers' code, and contribute to open-source initiatives.

Professional IT Organizations

- **ISC2** – This is the organization that boasts the CISSP exam. ISC2 has chapters across the US and encourages you to start one if there isn't a local chapter. They offer quite a few member benefits such as webinars, mentorship opportunities, test question writer, even a program to give back to small businesses.
- **CompTIA** – CompTIA is the issuing organization of many popular IT exams like Network+, Security+, CASP+. The benefit of these certifications is they are usually not vendor specific. This is a very common starting path for people wanting to break into the industry.
- There are so many organizations out there that target different specializations in IT. I highly recommend researching organizations in the area you have interest in. Maybe they have a local chapter or virtual meetings set up that you could join. The benefits and dues will vary per organization. It is important to get involved as that will expand your professional network, help you grow in the profession, and give you the opportunity to help others grow as well.

Cert Practice Test Resources

- **BOSON** – This one is my personal favorite. I have used them many times for multiple exams. In my opinion, the quality of the question banks is great. At the time of this writing, they offer a refund if you fail your exam after purchasing and using their practice tests. Other stipulations apply of course. They can be considered expensive, but are also called the gold standard in practice tests.
- **Udemy** – Udemy is excellent for getting a lot of practice tests cheap. They are usually good quality and created by good instructors as well.
- **Pocket Prep App** – I didn't stumble upon this until I started working on my CISSP. This is an app that you can customize the types of quizzes you take. You can take 1015 question quizzes whenever you have some free time. It will even let you know which domains you need to study a little

more in. It has many different certification question banks on there for one monthly fee. I wish I had known about it earlier on in my journey.

- **Instructor Websites** – It is common for some instructors to have video courses available on CYBRARY, Udemy, LinkedIn Learning, etc. but also have their own site. They will sell their courses there and can even have test banks available to get for the exams they instruct on. If you find an instructor you like learning from, check to see if they have a website. They may have learning material for other exams you have interest in taking or even have additional practice questions or beneficial resources for you on their own site.
- **Books** – In addition to being able to buy books for many of the certification exams, many also have books available specifically for practice tests. Occasionally they will be included in the main book. Look up what exam you need practice tests for and read reviews of people who have used it and taken the same test. If you use a book and find it helpful, take the time to write a review. This could help another person aspiring to pass the same exam when they see your review.

Health and Well-being

- Fitness is a big part of my life, especially with remote work. I can tell a difference in my mood and energy when I take breaks to walk throughout the day or take an hour to work out in my home gym. Everyone is different, but be mindful of what your body needs. There are so many free fitness apps and videos that you can access from home to do quick 10–15-minute workouts. There are even workout trackers to record your workout and remind you when it is time to get active.
- Yoga/stretching can be a great way to get moving. There are many apps tailored to each that can be done at home. This can be a great alternative to someone who may not have the time to go to the gym.

Networking and Professional Development

- **LinkedIn** – This is one of the go-to sites for professionals. It is a great platform to connect with people you meet at networking evens, conferences, or on the job. It is also a great place to make yourself stand out by posting relevant content for your desired audience. Be sure to follow popular professionals/influencers in the industry you are a part of or have interest in joining. Some will share lots of great content.

- **Discord** – Discord is a place popular for connection channels. There are channels for lots of things, especially IT related topics. Find some channels that interest you, then join. Ask questions, provide insight into topics you are knowledgeable about and don't be afraid to reach out to people directly. One new contact could change your whole career.

Cybersecurity Analysis Tools

- VirusTotal: A tool for analyzing suspicious files and URLs to detect types of malware. This can be used when you receive suspicious looking links via email or text. DON'T CLICK IT, copy and paste in VirusTotal.
- WHOIS Lookup: Provides domain and IP address lookup to identify the owner and operator of a domain.
- Wireshark: A network protocol analyzer used for network troubleshooting and analysis.
- Have I Been Pwned: Allows users to check if their personal data has been compromised in data breaches.
- Sucuri SiteCheck: Offers a free website security check and malware scanner.

Appendix C: Additional Resources & Links

Professional Organizations:

https://www.comptia.org/
https://www.isc2.org/
https://www.isaca.org/
https://www.issa.org/
https://www.ieee.org/
https://www.pmi.org/

Learning Resources:

https://www.cybrary.it/
https://www.youtube.com/
https://www.udemy.com/
https://www.pocketprep.com/
https://www.linkedin.com/learning
https://www.boson.com/
https://www.itpro.tv/
https://www.coursera.org/
https://www.hackthebox.com/
https://tryhackme.com/
https://www.pluralsight.com/

Online Forums and Communities:

https://github.com/
https://www.reddit.com/
https://discord.com/

Cybersecurity Tools:

https://sitecheck.sucuri.net/
https://www.virustotal.com/gui/home/upload
https://www.whois.com/
https://haveibeenpwned.com/
https://www.wireshark.org/

Password Managers:
https://bitwarden.com/
https://1password.com/
https://www.lastpass.com/

IT News/Updates:
https://feedly.com/
https://thehackernews.com/
https://www.bleepingcomputer.com/
https://krebsonsecurity.com/
https://www.darkreading.com/
https://www.itsecurityguru.org/

Other Resources:
Check out my personal website for other materials to help you along your journey. https://www.unitethegrid.com

Connect with me on my socials below:
Facebook: https://www.facebook.com/TylerLStrong

LinkedIn: https://www.linkedin.com/in/tylerlstrong/

X: https://twitter.com/tylerlstrong

Appendix D: Resume Examples

Professional Summary Examples

1. **IT Professional with 10+ Years Experience:** Seasoned IT professional with over 10 years of experience across various tech domains, including network administration, cybersecurity, and cloud computing. Proven track record of leading successful IT projects, optimizing network systems, and implementing robust security protocols. Skilled in managing cross-functional teams and driving innovation in IT solutions. Seeking a challenging role where extensive IT expertise can contribute to strategic technology initiatives and business growth.
2. **Non-IT Professional Transitioning to IT:** Dynamic professional with a rich background in [previous industry] and a recent pivot to the IT sector. Combines solid analytical skills and problem-solving abilities with a passion for technology. Recent training in [specific IT areas, e.g., software development, data analytics] and hands-on experience in [relevant IT projects or internships]. Eager to apply transferable skills and new IT knowledge in a stimulating environment that encourages continual learning and growth in the tech industry.
3. **Recent Graduate with IT Degree:** Newly graduated from [University Name] with a degree in [specific IT field], ready to launch a career in the IT industry. Adept in [list of IT skills learned during degree], with a keen interest in [specific IT area, e.g., AI, cloud computing]. Completed [number] of projects in [areas of study], showcasing skills in [list specific skills]. Seeking an entry-level position that allows for the application of academic learning in a practical setting and offers opportunities for professional development.

Mock Resumes

These mock resumes are fictional examples for illustrative purposes only

1. IT Professional with 10+ years of experience

John Doe 1234 Tech Lane, Silicon Valley, CA (123) 456-7890 | john.doe@email.com | LinkedIn: yourlinkedinprofilelink

Professional Summary Seasoned IT professional with over 10 years of experience in network administration, cybersecurity, and cloud computing. Demonstrated success in leading IT projects, optimizing network systems, and implementing security protocols. Adept in team management and technological innovation, seeking to contribute to strategic initiatives in a dynamic organization.

Professional Experience

Senior IT Manager XYZ Corp, Silicon Valley, CA | June 2016 – Present

- Lead a team of 15 IT professionals in managing the company's technology infrastructure.
- Spearheaded the migration of company data to a cloud-based system, improving efficiency by 30%.
- Developed and implemented a comprehensive cybersecurity strategy, reducing security breaches by 40%.

Network Administrator ABC Technologies, San Francisco, CA | May 2011 – May 2016

- Managed and maintained a network of over 200 workstations and 20 servers.
- Conducted regular network assessments to ensure optimal performance and security.
- Collaborated with the IT team to deploy new hardware and software solutions.

IT Support Technician Tech Solutions, San Jose, CA | July 2008 – April 2011

- Provided technical support to clients, resolving issues with hardware, software, and network systems.
- Assisted in the development and maintenance of the company's internal IT infrastructure.
- Conducted training sessions for staff on new technologies and best practices.

Education Bachelor of Science in Computer Science University of California, Berkeley | Graduated 2008

Certifications

- Certified Information Systems Security Professional (CISSP)
- Cisco Certified Network Associate (CCNA)
- Microsoft Certified Solutions Expert (MCSE)

Skills

- IT Project Management
- Network Security
- Cloud Computing
- Team Leadership
- Strategic Planning

Professional Memberships

- Member, Association for Computing Machinery (ACM)
- Member, IEEE Computer Society

2. Non-IT Professional Transitioning to IT

Jane Smith 789 Innovation Drive, Tech City, NY (123) 456-7890 | jane.smith@email.com | LinkedIn: yourlinkedinprofilelink

Professional Summary Dynamic professional transitioning from a successful 8-year career in marketing to the IT industry. Equipped with a passion for technology and recent certifications in network security and systems administration. Strong analytical, communication, and problem-solving skills, seeking to leverage marketing experience and newly acquired IT knowledge in a challenging IT role.

Professional Experience

Marketing Manager Creative Solutions Inc., Tech City, NY | June 2014 – Present

- Led a team of 10 in developing and executing marketing campaigns, increasing brand visibility by 40%.
- Collaborated with IT department to integrate new digital tools, enhancing campaign effectiveness.
- Managed budgets over $500k and provided data-driven strategies for marketing decisions.

Marketing Coordinator Innovative Marketing, Tech City, NY | July 2012 – May 2014

- Coordinated various marketing projects, liaising between clients and the creative team.
- Assisted in digital marketing efforts and website optimization for client engagement.

Education Bachelor of Arts in Marketing State University of New York | Graduated 2012

Certifications

- CompTIA Network+
- CompTIA Security+

Skills

- Project Management
- Digital Marketing
- Basic Network Security
- Systems Administration
- Cross-functional Collaboration

Professional Development

- Currently pursuing Cisco Certified Network Associate (CCNA)
- Completed various online courses in IT fundamentals and cybersecurity

Interests

- Member of Tech City IT Enthusiasts Club
- Volunteer at local community center teaching basic computer skills

3. Recent Graduate

Emily Rodriguez 456 University Avenue, Grad Town, MA (123) 456-7890 | emily.rodriguez@email.com | LinkedIn: yourlinkedinprofilelink

Professional Summary Recent graduate with a Bachelor's degree in Information Technology from Massachusetts Institute of Technology, enthusiastic about starting a career in the IT industry. Specialized in cloud computing and data analytics, with a strong foundation in programming and network security. Eager to apply academic knowledge in a practical, real-world IT environment.

Education Bachelor of Science in Information Technology
Massachusetts Institute of Technology | Graduated 2022

- Specialization: Cloud Computing and Data Analytics
- Senior Project: Developed a cloud-based data analytics platform for real-time data processing.

Internship Experience

IT Intern Tech Innovations, Boston, MA | June 2021 – August 2021

- Assisted in the maintenance and troubleshooting of network systems.
- Participated in a project to migrate company data to a cloud-based storage solution.

Skills

- Cloud Computing
- Data Analytics
- Programming (Python, Java)
- Network Security
- Problem Solving

Certifications

- AWS Certified Solutions Architect – Associate (In progress)
- CompTIA Security+ (Planned)

Academic Projects

- Cloud Data Analytics Platform: Developed a platform for processing and visualizing real-time data using AWS services.
- Network Security Analysis: Conducted a comprehensive security audit for a simulated network environment.

Professional Development

- Active participant in MIT's IT Club, focusing on emerging technologies and industry trends.
- Attendee of several tech conferences and seminars, including MIT's Annual Tech Symposium.

Appendix E: References

Trend Micro. (2016). 2012 LinkedIn Breach: 117 Million Emails and Passwords Stolen Not 6.5M. Retrieved from https://www.trendmicro.com/vinfo/us/security/news/cyber-attacks/2012-linkedin-breach-117-million-emails-and-passwords-stolen-not-6-5m

Wikipedia. (n.d.). WannaCry ransomware attack. Retrieved from https://en.wikipedia.org/wiki/WannaCry_ransomware_attack"

Wikipedia. (n.d.). Morris worm. Retrieved from https://en.wikipedia.org/wiki/Morris_worm

Wikipedia. (n.d.). ILOVEYOU. Retrieved from https://en.wikipedia.org/wiki/ILOVEYOU

Wikipedia contributors. (n.d.). Stuxnet. Wikipedia. Retrieved from https://en.wikipedia.org/wiki/Stuxnet.

Wikipedia contributors. (n.d.). 2017 Equifax data breach. Wikipedia. Retrieved from https://en.wikipedia.org/wiki/2017_Equifax_data_breach.

Krebs, B. (2023, December). Ten Years Later: New Clues in the Target Breach. Krebs on Security. Retrieved from https://krebsonsecurity.com/2023/12/ten-years-later-new-clues-in-the-target-breach/.

Wikipedia contributors. (n.d.). Facebook–Cambridge Analytica data scandal. Wikipedia. Retrieved from https://en.wikipedia.org/wiki/Facebook%E2%80%93Cambridge_Analytica_data_scandal

Made in the USA
Monee, IL
13 September 2026

62980082R00134